## About Island Press

Island Press is the only nonprofit organization in the United States whose principal purpose is the publication of books on environmental issues and natural resource management. We provide solutions-oriented information to professionals, public officials, business and community leaders, and concerned citizens who are shaping responses to environmental problems.

In 1998, Island Press celebrates its fourteenth anniversary as the leading provider of timely and practical books that take a multidisciplinary approach to critical environmental concerns. Our growing list of titles reflects our commitment to bringing the best of an expanding body of literature to the environmental community throughout North America and the world.

Support for Island Press is provided by The Jenifer Altman Foundation, The Bullitt Foundation, The Mary Flagler Cary Charitable Trust, The Nathan Cummings Foundation, The Geraldine R. Dodge Foundation, The Charles Engelhard Foundation, The Ford Foundation, The Vira I. Heinz Endowment, The W. Alton Jones Foundation, The John D. and Catherine T. MacArthur Foundation, The Andrew W. Mellon Foundation, The Charles Stewart Mott Foundation, The Curtis and Edith Munson Foundation, The National Fish and Wildlife Foundation, The National Science Foundation, The New-Land Foundation, The David and Lucile Packard Foundation, The Surdna Foundation, The Winslow Foundation, The Pew Charitable Trusts, and individual donors.

## About The New England Forestry Foundation

The New England Forestry Foundation is dedicated to the conservation and ecologically sound management of New England's private and municipal forest lands. The Foundation owns 20,000 acres in a system of 105 Foundation Forests. The Foundation carries out its work through programs in land conservation, Foundation Forest management and demonstration forest resource education, and particpation in formulation of regional forest policy.

# Legal Aspects of Owning and Managing Woodlands

# Legal Aspects of Owning and Managing Woodlands

Thom J. McEvoy

ISLAND PRESS
Washington, D.C. ♦ Covelo, California

The author expresses grateful acknowledgment for permission to use the following copyrighted material: Figures 4.1, 4.7, 4.8, and 4.9 are from *Real Estate Principles and Practices*, 8th edition, by A. A. Ring and J. Dasso, copyright © 1977. Reprinted by permission of Prentice-Hall, Inc., Upper Saddle River, NJ; figure 4.10 is reprinted by permission of Magellan Systems Corporation; figure 7.2 is reprinted by permission of the Society of American Foresters.

Library of Congress Cataloging-in-Publication Data
McEvoy, Thomas J.
Legal aspects of owning and managing woodlands / Thom J. McEvoy.
p. cm.
Includes bibliographical references and references and index.
ISBN 1–55963–638–6 (cloth : alk. paper). — ISBN 1–55963–639–4 (paper : alk. paper)
1. Forestry law and legislation—United States—Popular works.
2. Forest management—Popular works. I. Title
KF1750.Z9M38 1998 98-8075
346.7304'675—dc21 CIP

Printed on recycled, acid-free paper 

Manufactured in the United States of America
10 9 8 7 6 5 4 3 2 1

"It was only after we pondered on these things that we began to wonder who wrote the rules for progress."

—Aldo Leopold, *A Sand County Almanac*

"Everything should be made as simple as possible, but not simpler."

—Albert Einstein

# Contents

# Foreword

As a young forestry student I learned to view a forest only as a physical-biological entity, in hindsight more complex and dynamic than I could then understand. While I was aware of the social and economic aspects of forestry, they didn't seem relevant to my interests in wildlife, trees, and forests. It didn't take long on my first job, however, to realize that my understanding of forestry was sorely inadequate. Experience soon taught that the legal, financial, and social problems of managing the woodlands for which I was responsible were far more challenging than the biological aspects of forestry. Only after some three decades of practicing and teaching forestry have I come to see how inextricably interwoven are the human and natural contexts of forests, and how essential it is to understand those relationships to manage our woodlands for the goals and values we seek.

Most owners of small woodlands know that managing their lands is complicated by problems they did not anticipate when they decided to buy the land or begin active management on a previously ignored or inherited tract. One could find hundreds of books and pamphlets giving advice on such tasks as tree planting, silviculture, harvesting, and forest protection, but little was available to help anticipate and resolve the inevitable legal and financial problems of woodland management. As a result, unwise investments, legal mistakes, and poor financial planning have probably caused more woodland management failures than fires, insects, and inappropriate silviculture combined.

For many woodland owners, this book can make all the difference between success and failure when it comes to legal matters related to owning and managing their lands. Although it is not a substitute for the advice of a professional forester or a qualified attorney, it will make private owners far more savvy consumers of those services. In practical, everyday language Thom McEvoy explains the historical roots, legal principles, and technical details of property ownership and management that will enable woodland owners to ask the right questions of their attorneys, consulting foresters, and everyone they encounter in the complex task of owning and managing land. Knowing the right questions will prepare them to make better choices of partners, forestry and tax consultants, financial advisors, and attorneys, and better able to judge their answers. Little of the information in this book is new, certainly not to a seasoned extension forester like Thom McEvoy. What makes it special and unusually valuable is the way the information is organized and put into both historical and practical perspective, leading the reader to see how decisions in one aspect of management affects others over time. McEvoy explains, for example, how early decisions about methods of ownership and financial management affect long-term goals, tax liability, and estate planning. He sees the legal and financial aspects of woodland management in the same integrated way that modern foresters are coming to understand complex forest ecosystems: Everything is connected and interactive.

This book will help the woodland owner to understand and appreciate that the often confusing and seemingly burdensome aspects of the law affecting land ownership are rooted in history as surely as a forest is the result of decades of natural biological succession. And, as we are learning that every aspect of a natural system is critical to the whole, McEvoy leads us to see that our unique rights as private landowners in the United States rest on a complex legal system that balances our rights and obligations with those of the public at large—a system that has evolved and been tested over time. Put simply, the past is always prologue.

This simple truth came home vividly for me a few years ago when I was trying to advise some Russian foresters in Siberia who were charged with transferring state lands to private ownership. I wish I had had this book then, for their questions raised all the subjects that McEvoy so clearly explains: What rights do owners have? How are those rights granted and protected? How is the land marked on the ground? What can people do with the land? Can they change the uses, subdivide it, sell it, give it away to family, inherit it, mortgage it? Can they sell the trees or minerals and keep ownership of the land? Can they exclude others from using it, or use it in ways that destroys wildlife habitat? Who decides? How is government financed to do all this?

I found myself telling the history of the United States and how we created constitutions and such democratic institutions as courts and local governments to carry out those social contracts. I tried to explain how they would have to "invent" a system of surveys, deeds, and titles; town halls and methods of recording deeds and contracts; corporations; a court system; and even lawyers and tax collectors. They were overwhelmed with the realization that they were faced with a task that took the United States several hundred years to develop and refine by building upon centuries of English Common Law!

I reflected recently on that experience in Siberia as I sat beneath a venerable old oak on some land we own along Lake Champlain. That tree has witnessed all those changes in its lifetime. As a seedling it was within the common land of the Abenaki tribe. When a century old, the old oak came within the boundaries of a parcel of private land defined in fee simple under English Common Law. It now stands on a tract of private land governed by the complex laws of our nation, state, and the local town, which both limit and protect the bundle of rights we hold in fee simple. Tracing the history of all the legal and financial claims on this tree is as much a part of understanding woodland management as understanding its ecological history as a red oak. In many ways, it is as daunting a task for us landowners and foresters here in the United States as it for those in Siberia seeking a way to create private land ownership. We are fortunate to now have Thom McEvoy's book to guide us.

*Carl Reidel*
*Professor of Forestry, Natural Resources,*
*and Environmental Policy*
*University of Vermont*

# Preface and Acknowledgments

I began my career in academics many years ago as a pre–law student at Michigan State University. Fascinated with law, and assured by family and friends that lawyers make a good living, it seemed like the perfect path. That is, until I took a course in forest ecology. Even before the end of my first year in college I had changed disciplines, and four years later—jobless—I did not regret it. Aside from a short stint as a consulting forester struggling to make a living in the aspen forests of northern Michigan, followed by graduate study in forest biology, I have spent my entire career as an extension educator in forestry, a job I love.

About fifteen years ago, I considered going back to school for a Juris Doctorate degree. So serious were my plans, I took the Law School Admissions Test and scored well enough to receive an offer for admission to the law school I wanted to attend. But events and lack of financing conspired against me and I never went. Instead, I audited an undergraduate course in business law at the University of Vermont. It was an excellent course, with one of the best teachers I have ever encountered. I was so inspired by the experience, I began to write and speak about legal aspects of owning and managing woodlands.

In 1986, I was invited to give a presentation to a group of nonresident woodland owners on the subject of selling timber. After the talk, one of participants—a lawyer—came forward to offer his compliments. Although there were several valid criticisms he could have shared, instead he applauded my unfettered and clear explanations of abstract and often difficult legal concepts. Little did he know it does not take much encouragement from the right sources to launch me headlong into a project. For more than ten years I have wanted to write this book. At first the task was daunting, a good reason to put it off. But I was haunted. The question from colleagues I had not seen in a while became: "So, when is the book coming out?"

An outline for the book began to take shape in the fall of 1995. I shared my ideas with anyone who would listen—woodland owners, foresters, loggers, lawyers, accountants, estate planners, and others. A year later, the outline became the proposal for a twelve-month sabbatical leave from my post as extension forester at the University of Vermont. In the fall of 1996, I circulated a draft table of contents for the book to my extension colleagues around the country and received from them much encouragement and a pile of materials on local regulations and state extension publications dealing with legal aspects of owning land. Some of those materials were very useful.

When I set to writing in December of 1996, the task proved every bit as daunting as I feared. I quickly became bogged down in details, realizing all but the most dedicated reader would fall fast asleep after the first few pages. It took weeks to locate a proper voice and to discover a formula that would allow me to cover a huge subject area in a small book. Throughout, I have taken the position that it is more important to know how to ask the right questions than to have answers to the wrong questions. Detail-oriented people will find the book unsettling; after all, how can anyone explain the complexities of, for instance, land trusts in just a few pages? Generalists will say, "Good, it is about time the emphasis is placed on broad understanding rather than mind-numbing details that make even the most well-accepted, broadly applied set of facts nothing more than fanciful wanderings of the author." In other words, it was pointless to write something that people would not read.

My goal was to write a book—that people will read—on a subject that is of great importance to woodland owners: understanding the legalities of real property, the rights and responsibilities as well as the opportunities and pitfalls of owning and managing forests. The prospects for future forests, and the importance of thinking like a forest when it comes to long-range planning, are also major themes. My measure of success in this endeavor is the degree to which readers are empowered to become more active forest managers.

I am not a lawyer. Thus, nothing in the text should be construed as legal advice or legal opinion. This book is based on information that is generally known and on my ideas, some of which have been inspired by other cited sources. The ideas are intended to provoke readers to consider strategies and methods to improve the long-term management and care of woodlands, and to seek the advice and guidance of qualified local professionals. Nothing in the book should serve as the sole basis for a reader's actions or inactions. For these reasons, the author and publisher disclaim responsibility or liability for loss that may result from a reader's interpretations or applications of the material in the book.

Many people deserve thanks for their help. Among them are two well-seasoned attorneys who reviewed the manuscript from cover to cover, offering many important and useful suggestions. They wish to remain anonymous "because of potential disagreements with colleagues." Because I know they will see the book, I want to express my gratitude here for their extraordinary efforts. Many thanks to Larry Bruce, a practicing attorney in St. Albans, Vermont, who read the manuscript and made many useful suggestions and who also agreed to be acknowledged.

Special thanks goes to Doug Hart, a woodland owner in Quechee, Vermont. Doug reviewed chapters throughout the year, and he gave the final manuscript an especially close reading. We communicated exclusively by e-mail, and his many excellent comments and suggestions have made my writing much easier to follow. The illustrations are the work of Marna Grove of Pencil Mill Graphics in Castleton, Vermont. Her creations were a truly collaborative effort, and it was a pleasure—as always—to work with her.

Others who deserve thanks include: Tami Bass, a surveyor in Vermont, for reviewing the chapter on surveys and boundaries and offering some tips on illustrations to help explain blazing; Sarah Tischler, attorney and estate planner, who was of immeasurable assistance in the preparation of the chapter on estate planning; Lloyd Casey, forest taxation specialist with the U.S. Forest Service, who reviewed the chapter on forest taxation and served as my primary source for information on the Taxpayer Relief Act of 1997; and Keith Ross, a forester with the New England Forestry Foundation, for suggesting the section on like-kind exchanges and for providing most of the information presented there. I also want to acknowledge the efforts of other reviewers, each of whom read the entire manuscript and offered comments to improve it: Harry Chandler, Stanley James, Farley Brown, and Extension Forestry colleagues Larry Biles and Rick Fletcher. Thanks also to the following people for help with various aspects of the project: Deborah LaRiviere, John McClain, Maria Valverde, and my eight-year-old son, Christopher, who helped organize one of the appendices. Finally, I want to thank my colleagues at the University of Vermont for allowing me a sabbatical leave to complete the project.

I wish to dedicate this book to two important women in my life: my deceased mother, Doris, who always told me I could do anything I put my mind to, even though I eventually learned this was not completely true; and my wife, Hallie, who makes each day a gift. Forever and always. . . .

*Thom J. McEvoy*
*Bolton Valley, Vermont*
*9 January 1998*

## Chapter 1

# Forestry: Past, Present, and Future

One of the strengths of forests in the United States is the diversity of people who own them. Of the 737 million acres of woodlands in this country, 58 percent is owned by families, individuals, estates, and businesses—a very high percentage by global standards. Private ownership is a strength because different owners have dissimilar ideas about forest resource values and how best to manage lands. They have disparate management schedules and different priorities. This leads to diverse landscapes, and forest diversity promotes healthy forest ecosystems, appealing vistas, and a steady supply of such forest benefits as timber, wildlife, and recreation.

There are 9.9 million private woodland owners in the United States, most of whom reside east of the Mississippi River (Sharpe et al. 1995). A disproportionately large share of the private forest acres (40 percent) are owned by a small number of corporations (6 percent), but the balance is owned by private individuals, called "Non-Industrial Private Forest" (NIPF) owners (Birch 1994). NIPF owners include any noncorporate holding of forests, such as individuals, married couples, partnerships, trusts, and other types of holdings.

With so many different private owners spread across 24 degrees of temperate latitude, one would expect a high level of diversity of goals and needs. All private forest owners have one thing in common, though: at some point during their forest tenure, they will encounter legal matters that are directly related to owning and managing woodlands. For most of them, the encounters will be simple—a sale closing, title search, or timber sale contract. Some will experience encounters with the law that are less simple—boundary disputes, liability, and conflicts of interest. The purpose of this book is to describe the legal nature of forests, from our ideas of private property, to the ethics of forestry practice, to the difficulties of planning for the future of forests we leave to our children.

This is not a definitive guide on how to deal with the simple and not-so-simple legal aspects of owning and managing woodlands, for obvious reasons: laws and forest-use traditions vary from place to place, and the circumstances surrounding one individual almost never apply to others. Because laws help define acceptable and unacceptable behavior, it is the application of law to specific behaviors or circumstances that makes each case different. There are many areas, however, where laws apply broadly to the experiences encountered by forest owners irrespective of where they live, and that is the emphasis of the chapters that follow.

Owning and managing forest lands is a legal endeavor; from forest acquisition to deeds and boundaries, from timber sales to bequests, a forest owner can do little without confronting laws, rules, regulations, and traditions that define acceptable and unacceptable practices. Aside from a general notion of when legal assistance is necessary, most forest owners do not fully appreciate the extent to which laws and regulations control or influence decisions. Too often a crash course in local laws follows a disagreement, and one can only hope to be on the right side.

The following chapters describe legal matters that are among the most common requests of public agencies that field inquiries from woodland owners. At the top of the list are questions about timber sales and dealings with foresters and loggers. Although the type of information a public forester might offer in a particular instance would probably not qualify as "legal" advice—at least by standards of the legal profession—it is often guidance an owner will listen to and follow. The difference between legal advice and the advice of a public forester is the degree to which information is tailored to a client's request. It is the wise forester who answers specific questions in general terms, with the objective of educating the owner. Specific advice—what to do and what not to do in a particular instance—is dangerous ground for a public servant.

There is no better measure of a democratic society's evolution than the laws by which it lives. The older our society, the more complex its laws become, because our legal system is intended to be adaptive, to change with society. The U.S. Congress and state legislatures make new laws that, in theory, reflect society's wishes. Decisions rendered in courts through the application of existing laws become "precedents" for subsequent decisions. The specific application of laws, precedents, and the procedures used to argue points of law in court are the main reasons we have lawyers. Not many lawyers, though, specialize in legal matters related to owning and managing woodlands.

Another reason for lawyers: The technical language of law is Latin—a "dead" language not in use by society and not apt to change. The law is peppered with Latin words and terms because they can have only one meaning. Ironically, the very thing that assures clarity to lawyers is usually a source of confusion for the lay person. Lawyer jargon, "legalese," is virtually incomprehensible unless you have studied law (or Latin). Some skeptics suggest lawyers use legalese to further society's need for lawyers. It is impossible to describe legal aspects of forests without some legalese, but my goal has been to keep it to a minimum.

The chapters that follow contain descriptions of a wide range of legal issues related to forest ownership and use. Any one of the chapters could be expanded into its own book. For the reader who is well informed in any one of the subject areas, it will be easy to see that depth was the victim of breadth.

Since the legal norm and traditions that affect forest ownership and management decisions are based on experiences and trends, a brief overview of forestry past, present, and future is in order. The history of forest use and society's response to the effects of timber harvesting on the intrinsic values of forests have complicated forest ownership. Although the rules and regulations we live by today often seem excessive and unreasonable, put in the perspective of the past they make more sense.

## Forestry Past

A forest is a collection of plants and animals that live in places where there is adequate rainfall (or nearby sources of water), sufficient soil for nutrients and anchoring, and favorable climatic conditions. More than half the land area of the United States meets the minimum environmental conditions for forests, and about one-third of U.S. soil currently supports forests. The difference between potential forests and existing forests is that potential forests are largely lands that have been cleared for agriculture and for human settlement. A broad swath of land west of the Mississippi River and east of the Rocky Mountains, however, is too dry to support forests, except along rivers and stream beds. West of that region, the presence or absence of forests is mostly related to elevation, aspect, and prior land use. Where there are mountains tall enough to force uplifting and cooling of prevailing winds, there is usually rainfall ample enough to support forests.

Forests are synonymous with trees, the predominant and most obvious inhabitant. It was land clearing for agriculture and the use and exploitation of trees for lumber and charcoal—from the eastern states to the Midwest, then on to the Far West—that led to westward expansion and settlement. With few exceptions, the history of forest use in this country is not pretty. Forests were felled as fast as prevailing technologies would allow, with little or no regard for future generations. In the early nineteenth century, forests seemed limitless, but by the beginning of the next century, almost all easily accessible forests had been harvested at least once. When the end of timber was in sight, the science of forestry was imported from northern Europe.

Modern forestry methods, with the goals of increasing harvesting efficiency and sustaining wood yields, were not widely accepted until after World War II. Government and wood-using industries played a leading role in promoting the cause of good forestry. From the late 1940s until the early 1970s, the message from a rapidly growing forestry community was this: "Forests are renewable." Trees were harvested in such a way that new forests were regenerated, or the site was replanted after harvest. The emphasis, though, was on wood yield.

In the mid-1970s, the message shifted to multiple use, implying forests can be easily managed for more than one resource at a time. Although "multiple purpose" might have better characterized the philosophy, managers were acknowledging competing forest values and beginning to experiment with ways to harvest timber while protecting or enhancing other forest values.

In the mid-1980s, studies of woodland owner attitudes and motivations revealed that many people were more interested in managing lands for wildlife and other intangible values than for timber. Forest owners were not opposed to harvesting timber. Rather, the motivation to harvest increased in direct proportion to the nonmonetary values of a particular action. For instance, an owner opposed to harvesting trees for the sake of income production will harvest to improve access or to create hiking and ski trails. Providing better habitat for wildlife is high on the list for many owners.

"Forest stewardship" was the theme du jour, and for the first time managers were developing forest-management plans where timber might be subordinate to other values. Forestry was rapidly evolving from an anthropocentric view of forests to a biocentric view.

## Forestry Present

Today, the principles of forest management in this country are still less than four generations old, sufficient time to regrow only an insignificant fraction of the presettlement forests. And yet the United States holds some of the most productive and well-managed forests in the world. Managers have learned a hard lesson from the past: Forests are much more than trees. They are an enchantingly complex association of plants and animals—including humans—that can provide many different combinations of benefits, of which timber is only one.

In the early 1990s, amid a highly divisive debate about forestry's responsibilities to threatened and endangered species, the concepts of biodiversity and ecosystem management emerged. Although the particulars of ecosystem management are discussed in greater detail in chapter 2, its emphasis is on principles and practices that mimic natural processes and thereby protect the integrity of forest ecosystems. A more appropriate term might be "ecosystem approaches to management," since an ecosystem, by definition, is self-sustaining and does not require management or human interventions. The term *management* implies human use, and a fundamental agreement of all who have joined the debate is that humans are part of forest ecosystems, and our use of forests must be factored into the picture.

In just the past ten years, forestry practice has experienced a revolution, the emphasis shifting from timber production to protecting forest ecosystems. Although the cause is not clear for this radical departure from traditional forestry with its emphasis on wood production, a heightened awareness of forest values among the general public, coupled with new science, has changed the way we operate.

Not every manager agrees. There are those who maintain a good forester always considers effects on whole ecosystems. Others charge that traditional forest-management approaches put foresters in the pockets of the wood-using industry, with little regard for nontimber values and the interests of future generations. Although the chances are slim of one side convincing the other it is wrong, both sides appear to want essentially the same thing: productive forests that provide a wide variety of benefits while protecting species and other nonmonetary values that make forests special. The most significant and important aspect of the debate is that it is happening at all. Forestry in the United States is evolving from paramilitaristic origins (possibly due in part to its Teutonic influences), where authority is never questioned and much is accepted on faith, to a highly multidisciplinary field, where there are few easy answers and many, many choices.

At the heart of the current debate is this: Can we sustain forestry production at current levels and still protect the productive capacity of forests, maintain or improve water quality, provide appealing landscapes and recreation, *and* provide adequate habitat for a diversity of forest-dwelling plants and animals? The traditionalists say, "That's what good forestry is all about." Others say, "Wood production is subordinate to protecting the long-term integrity and health of the forest. If we take care of the forest, wood will be available but probably not at rates thought reasonable just a few years ago."

The concept of "sustainable forestry" is gradually replacing "sustained yield" as the primary goal of good forest management. If a practice is not sustainable—ecologically, economically, and socially—then the practice should be avoided. Wood yield by itself is no longer an adequate measure of good forestry. For some managers—even those with college textbooks no older than the mid-1970s—weaned on "sustained yield," sustainable forestry is a radically new idea.

The forest industry, too, is changing. Through its Sustainable Forestry Initiative (SFI), the American Forest and Paper Association, representing many wood-using companies in the United States, is trying to encourage its members to adopt "guidelines" intended to protect forest ecosystems. Among the guidelines are new procurement policies aimed at keeping loggers out of the woods when forests are especially susceptible to damage, commitments to quickly regenerate harvested lands, and training programs for loggers. The goal of SFI is to shift the emphasis from wood production to long-term forest management. Because many companies are publicly held and have a fiduciary responsibility to shareholders, any change that affects the bottom line is risky. For a risk-adverse industry not known for rapid change, the SFI initiative is a valid attempt to move in the direction of more sustainable forestry. As long as short-term profitability is a motive, however (as it is for any company with shareholders), there are limits to which SFI principles and guidelines can be implemented. Only when there is a short-term financial advantage to practicing sustainable methods will the incentive exist to do so.

## Forestry Future

Because ecosystems are nested (i.e., smaller ecosystems, such as a population of lichens on the trunk of a tree, are part of larger systems, the largest of which is the earth), protecting ecosystem values requires societal decisions that extend far beyond individual owners, local communities, state legislatures, and Congress (figure 1.1). True ecosystem approaches to forest management must transcend

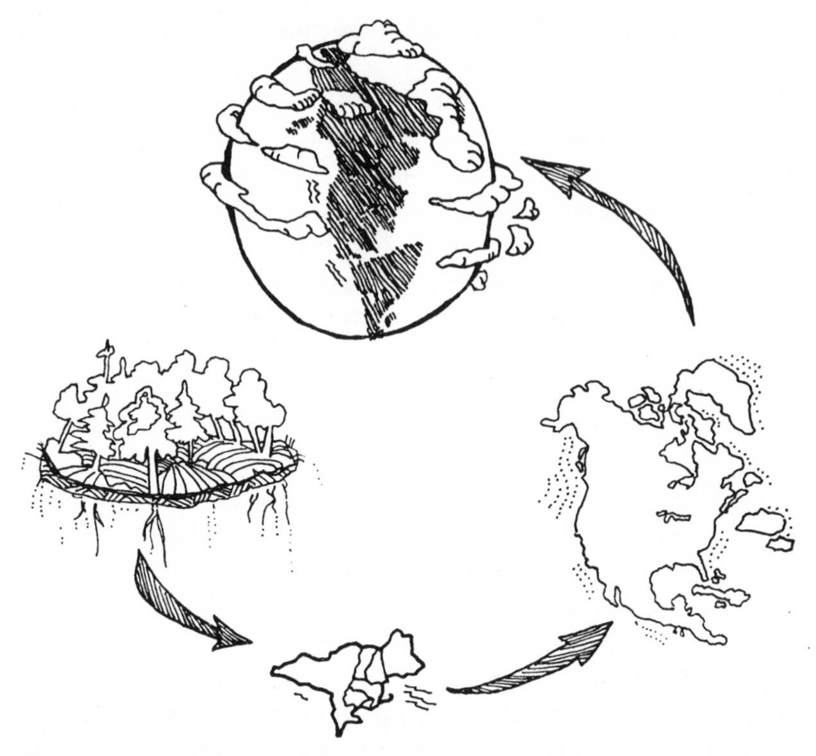

*Figure 1.1. An ecosystem can be defined at many different levels, from the lichens growing on the bark of a single tree, to entire landscapes, continents and—ultimately—the earth. The hierarchical effect is what makes ecosystems so complex, enchanting, and challenging.*

geographic boundaries through multilateral partnerships, requiring an unprecedented level of cooperation across many borders—political, geographical, and disciplinary. Policies to protect ecosystem values may thwart a particular owner's management plans. Herein lies the future forestry debate: How can we protect whole ecosystems without impinging on private property rights? Who makes decisions about which ecosystem values are critical and how we protect them? Who is going to pay for ecosystem management if it costs more than traditional methods? There are no easy answers, and some people suggest that questions such as these are the Achilles' heel of ecosystem management. Almost everyone, though, despite where they stand on the issues, agrees: Ecosystem approaches to managing forests are here to stay.

While some people are daunted by the prospects of implementing forest ecosystem–friendly policies and practices, others see opportunity. There are today no firm guidelines on the techniques and methods of managing whole ecosystems, but innovative forestry professionals are committed to "adaptive management," based on the theory that when in doubt, it is better to err on the side of inaction, knowing that practices will change as new science becomes available. Woodland owners are increasingly receptive to new practices, even if it means less immediate income or longer investment periods.

An emerging trend in the marketplace may make the wait worthwhile. Consumers in Europe are favoring products that are manufactured in an environmentally sensitive fashion. To give consumers credible product pedigree guarantees, the manufacturing process is monitored and certified by independent, third-party organizations. The certification process extends from raw material to shelf, by maintaining a chain-of-custody. This gives the consumer assurances a product has been manufactured without unacceptable exploitation of people or the earth. Also known as "green certification," it is an attempt to inform consumers of the environmental and social consequences associated with the products they use. The idea is still relatively new in this country, and there is no reliable evidence to suggest consumers in the United States would seek out or pay more for green-certified products. Some manufacturers and retailers, though, are demanding green-certified raw materials or final products, and there is a growing trend to market the earth-friendly aspects of those products. It may be only a matter of time before consumers favor green-certified products over those of unknown origin. As consumer demand for certified products increases, the demand for certified timber will grow. Forest owners who are aware of this trend, and agree to abide by the principles of certification, will be in a good position to supply the demand.

Product certification is slowly catching on among primary wood producers, foresters, and woodland owners. A woodland property (or manager of more than one holding, in some circumstances) becomes certified when it is clear to a certifying organization that the owner is adhering to a set of principles that includes ecosystem approaches to management. The principles also include a commitment to sustain wood production, to protect wildlife habitats, forest streams, and scenic landscapes, and to utilize the values of forests in a nonexploitive fashion. Because certified wood is in short supply, a buyer (theoretically) will offer a premium for wood that can be sold to customers as "green-certified." The idea is similar to a grade stamp on lumber, although the certification has nothing to do with wood quality but only with how the wood was produced.

Most people involved in wood product certification see demand building from the consumer back to the woodland owner. It may take years for this to develop in the United States. The experiences with consumers in Europe, however, suggest interest in green certification will increase among U.S. consumers. For some woodland owners, who are committed to the principles of managing whole ecosystems and are willing to wait for markets, product certification (both the cost of certification and the added expense of longer rotations) may be a sound investment.

Even though society's expectations of woodland owners are changing—from providing forests as a setting or background, to making them everybody's backyard—owning and managing forest lands is still a winning proposition for many families. The future holds promise for all woodland owners, but it is especially promising for those who make long-term commitments to the land, commitments that extend beyond a lifetime. It is this need for continuity from one generation to the next that both strengthens families and ensures healthy and productive forests for the future. The days of passive partnerships between woodland owners and the forestry community are over. The future forest will require a much higher level of participation, and helping prepare woodland owners to engage the challenge is the goal of this book.

# Chapter 2

# Private Property

In the United States, the concept of the ownership of land and its natural resources has an interesting history, rooted in the feudal system developed by the early monarchs of England. Initially, all land was vested in the king by divine right. But because it is impossible for one person to effectively manage and control such vast areas, the land was divided among favored lords, who acted as agents of the king. Lords further subdivided the lands into tracts managed by vassals—the people who planted crops and raised stock—in exchange for a share of their production and for a commitment to serve the king in times of war. The lord was granted the privilege of making the king's lands productive, ensuring adequate revenue for the monarch and a steady supply of men willing to defend (or expand) the king's claims. Generally, a lord's claim could be passed within the family by inheritance but could not be sold. Vassals had no rights, although sons and daughters usually followed a parent's tenure at the pleasure of the lord. The king always retained the rights to timber and wildlife of the forest, and use was not allowed without his express permission.

In time, monarchs discovered other ways to ensure a supply of men for their armies, and gradually property interests were vested in individuals without any controlling interest by the king, except the king's rights to collect taxes. Known as the "allodial system of property," it is the basis for the common law of land titles in most of the United States. The allodial system recognizes an individual's rights to land that extends from the "center of the earth to the heavens."

When North America was settled by England, France, Spain, and Holland, the monarchs of those countries granted colonial charters and grants that gave individuals the rights to the soil (and the timber with some exceptions) but not the rights to minerals or to establish political boundaries. Land was essentially free to the common man who could clear, settle, and record a claim to it, so long as it was not part of an existing grant. As the best lands had already been claimed, immigrants were forced to look westward. From England's perspective, by 1774 the situation was getting out of hand. In that year, the English Parliament passed the Quebec Act, which attempted to lay claim to all lands west of the colonies, as far south as the Ohio River. Any person wishing to settle beyond the Alleghenies required permission from the king. Individuals who viewed the colonies as possessions of England were restricted in their westward movement. Those who were doubtful of the king's claims ignored his edicts, setting the stage for the Revolutionary War. Significantly, the Declaration of Independence was mostly over land and the colonists' claims that the king was meddling in the growth and exploitation of vast new areas.

After the Revolutionary War, much of the business of Congress focused on land. With more land than cash after the war, it was easier for Congress to pay soldiers with grants of land, mostly west of the Appalachian Mountains. Each veteran received a grant of 320 acres, and many sold their grants to speculators without ever having seen the land (Ambrose 1996). One of the first acts of the new Congress (Resolution of Congress on Public Lands, 1780) vested in the federal government rights to all lands west of the colonies by trade, purchase, treaty, or conquest. History books have led us to presume Congress adequately and fairly resolved all aboriginal claims to the lands of Native Americans, but most of North America is the spoils of conquest. Original grants and even the most primitive abstract of title do not recognize the original claims of Native Americans, and yet rivers, mountains, and vast regions still retain Native American names for those areas (figure 2.1). By 1862, nine acts of Congress dealing specifically with land had been passed, leading the way to real property ownership as we know it today.

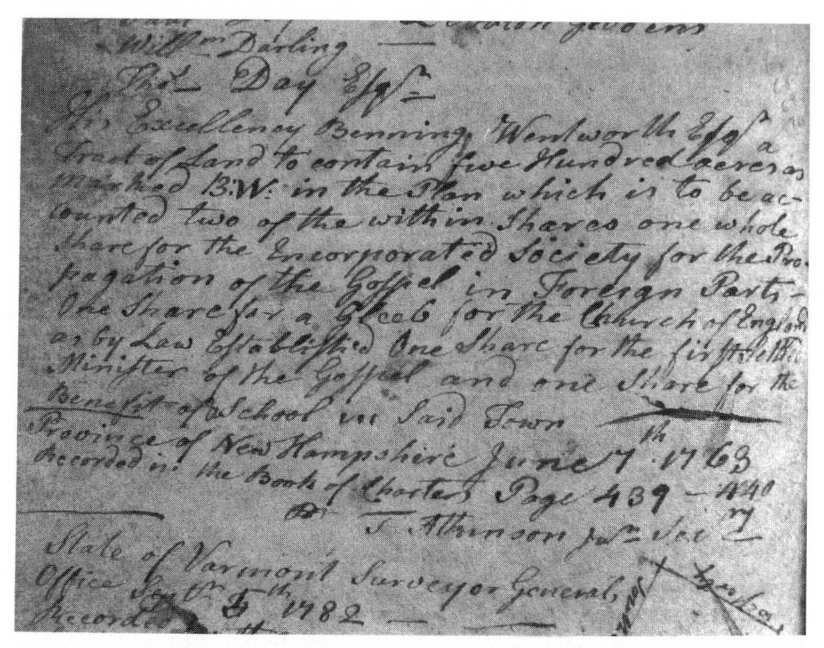
Will^m Darling —
Tho^s Day Esq^r
His Excellency Benning Wentworth Esq^r a
Tract of Land to contain five Hundred acres as
marked B:W: in the Plan which is to be ac-
counted two of the within Shares one whole
Share for the Incorporated Society for the Pro-
pagation of the Gospel in Foreign Parts -
One Share for a Glebe for the Church of England
as by Law Established One Share for the first settled
Minister of the Gospel and one Share for the
Benefit of a School in Said Town
Province of New Hampshire June 7th 1763
Recorded in the Book of Charters Page 439 — 440
T. Atkinson Jun^r Sec^ry
State of Vermont Surveyor General's
Office Sept^r 5th 1782
Recorded

*Figure 2.1. The original grants for the town of Bolton, Vermont, in 1768.*

## The Bundle of Rights

Real property differs from personal property in one very important and significant way: It cannot be moved from place to place. The only interests one can hold in real estate are the rights associated with ownership. These include the following:

1. The right to possession or the right to establish boundaries that signify to others that land belongs to you.
2. The right to control how the land is used, whether for crops or timber, pasture or open-pit mine.
3. The right to enjoy the benefits of land, including any income it may be capable of producing, or the intangible benefits from the aesthetics of land.
4. The right to give, sell, encumber, or bequeath your rights—or a portion of your rights—to others.

These four rights are necessarily broad and represent the connection to common law ideas of allodial property rights we inherited from England.

The bundle of rights currently recognized in U.S. jurisprudence as they pertain to private ownership of real property is an expansion, or restatement, of the English common law rights associated with ownership. They include the rights of use, occupancy, cultivation, exploration, the right to minerals (and the extraction of minerals), the right to sell or assign interests in the land, the right to license or lease, the right to develop, the right to devise and inherit, the right to dedicate and give away and to share, the right to mortgage and exercise a lien, and the right to trade or exchange land. (Pearson and Litka 1980). However—and this is a key point, the grist of much of the debate about private real property ownership rights—the exercise of these rights is subject to limitations the state may impose for the sake of protecting public interests. An owner's rights are not absolute and never have been.

In exchange for the state's willingness to defend an owner's just claims in land, the state also has rights associated with private lands. These include the right to tax land, the right to take land for public use with just compensation (eminent domain), the right to control use to ensure protection of the public's interests, and the right to claim title to the land when an owner dies intestate and there are no legal heirs to the property (escheat). The state also lays claim to wildlife that inhabit private lands. An owner has the right to hunt or trap, provided he or she does so with a license and during the appropriate season established by the state. An owner can choose to lease or sell to others the right to hunt, with the same state-imposed conditions, but wildlife are free to roam regardless of property boundaries. The private owner has rights to protect land from wildlife depredations by fencing or other means, such as to protect crops, but does not own the animals.

When society extends rights to individuals, it tacitly expects the individual to accept certain responsibilities. It is assumed an owner is not "sleeping on his rights" and that the owner is fully aware and in control of activities on the property at all times. This means that if the owner sells timber to a logger who proceeds to violate the state's Clean Water statutes, the owner is responsible for the violation. (Sharing liability for upholding a state's laws, especially during a timber sale, is discussed in more detail in chapter 6.) As a forest owner you must understand the extent of your liability for the careless actions of others. (Personal liability is discussed in chapter 3.)

## Private Rights versus Public Interests

Deciding when the public's interests are at stake, especially in light of the fact that the costs of ownership—taxes and other expenses—are borne by the individual, is often a divisive and heated debate. The bottom line is, a forest owner must adhere to the laws, rules, and regulations controlling forest use and avoid doing anything that is apt to have a substantially negative effect on neighbors, the community, or society.

During more than two hundred years of U.S. real property law, broad rights guaranteed by the Constitution and defended by the state have become more narrowly defined. Today, the rights we hold in real property spring from society (Barlowe 1990), and society's interests may be far different from those of an individual. Property owners decry the erosion of their sovereign rights, while environmental groups and others say it is time to shift the debate from rights to responsibilities. The courts are in the middle of the debate, but it appears evident the days of a laissez-faire approach to the rights of private landowners are nearly over.

When a real property owner's rights are lost by government action during his or her tenure, it is called a "taking." The taking of rights is preceded by a condemnation promulgated by a governmental entity that is staking a claim against your rights for the public good. For instance, when the county decides to build a new road that crosses your property, it will condemn the current uses of the land and take it (by right of eminent domain) for the new highway.

Any involuntary loss of real property rights is a taking. The Fifth Amendment of the U.S. Constitution requires the taking government to pay "just compensation." But the authors of the amendment were thinking of eminent domain, where the state takes all rights. They failed to consider circumstances where regulations, statutes, and rules might restrict some but not all rights. Also, what constitutes just compensation is rarely agreed to easily, since it usually has nothing to do with fair market value. For instance, if hay land is taken for a new road, "just compensation" is apt to equal the capitalized cost of hay to the farmer even though the land could have been developed into housing or some other higher fair market value. Recently, the Supreme Court has held (through application of the Fifth Amendment) that if the state's (or community's) actions represent a taking of some of a property owner's rights, the owner should be compensated for those rights—assuming the restrictions, or taking, happened during the owner's tenure and not before, when he could have or should have known his rights were restricted. There must also be an actual loss of economic value (*Lucas v. South Carolina Coastal Commission*, 1992). The high court's position is that a taking occurs when there is a substantial economic loss through the direct actions of a governmental authority after the owner has acquired property. "Substantial economic loss" in the South Carolina case was more than 90 percent of the purchase price of the property. The South Carolina Coastal Commission had instituted a restrictive law after Lucas acquired his barrier-island property but before a devastating hurricane. The law says hurricane-destroyed houses on barrier islands may not be rebuilt. The law constituted a "taking" because it was instituted after Lucas built on the property, and it had a major effect on fair market value. Although he still owned the land, he could not use it for its intended purpose: a beach house.

How then does the state compensate for a partial loss of rights, and what is a reasonable threshold for loss beyond which the owner should be compensated? When is a taking capricious and when is it necessary or unavoidable? To what extent should society be responsible for the risks associated with the acquisition of any property, whether stocks and bonds or forest land? Forest property in urbanizing areas will be subject to an increasing level of control by municipalities. From zoning to restrictions on forest practices, forest owners can expect to see more control, not less, over how they manage their lands.

In 1995 alone more than one hundred bills containing "partial takings" provisions were introduced in thirty-nine states (Zhang 1996). Already, eighteen states have attempted to protect private property rights from government regulatory takings by enacting laws that change the assessment on these properties, provide compensation for losses, or a combination of both. (Ibid.) States with updated statutes dealing with this subject include Arizona, Delaware, Florida, Idaho, Indiana, Kansas, Louisiana, Minnesota, Mississippi, Missouri, North Dakota, Tennessee, Texas, Utah, Virginia, Washington, West Virginia, and Wyoming. Information about these statutes can be obtained from the applicable Attorney General's office (Appendix B).

Problems with regulatory taking statues arise when properties are appraised to determine loss. There are as many different approaches to forest land appraisal as there are appraisers, and there are no uniform guidelines. Most often, loss is calculated as compensation for preventing a current use, not for loss of fair market value.

Recent state interest in property rights will force Congress to pass federal laws that provide guidance and consistency to the states. Property rights advocates are constantly spurring action in Congress. The House of Representatives passed a bill that allows compensation for loss if 20 percent or more of the value of the property is "taken" (and the owner keeps the property). If the loss exceeds 50 percent, the owner is fully compensated. To qualify, the loss must have resulted from the Clean Water Act, the Endangered Species Act (ESA) (both are discussed in more detail below), or agricultural conservation compliance under the Food Security Act. A Senate version establishes a 33 percent threshold for compensation. Neither of these bills is expected to become law in its present form, but the debate about property rights will continue. Until Congress acts, courts will validate a taking if there is a valid public purpose, including a "real and substantial threat to public health." Compensation may be awarded if the taking has caused an "inordinate burden" on the owner.

Before undertaking a forestry project in your woodlands, discover from knowledgeable sources any local rules or regulations that may apply to your activities. A good source for this type of information is your local state extension forester (Appendix A). Local government regulation of private forestry practices has increased more than fourfold in the past ten years (Martus et al.

1995). Generally, local rules and regulations are easy to comply with even though they may add a little extra nuisance.

Also, if you have concerns about local regulations, become involved in the debate. Remember the prophetic words of West Coast logger Bruce Vincent, "The world is run by those who show up."

## Protecting the Public's Interests

You buy land, survey the boundaries, inventory the resources, design access routes, develop a long-term management plan, and pay taxes. Why then should anyone have the right to tell you what to do on your land or how to do it? After all, so long as your activities do not have a negative consequence for your neighbors, you have the right to use the land as you see fit. Right? Wrong! The public has an interest in your land, and in recent years the U.S. Congress, state legislatures, and local municipalities have passed laws, regulations, and rules that you must uphold to avoid fines and other penalties. The bundle of rights is not nearly as sacrosanct as it used to be—from society's perspective, you share certain rights with the public.

Two acts of Congress have had a substantial bearing on forest-management activities: The Endangered Species Act (1973) and the Clean Water Act (1977).

### *The Endangered Species Act*

The Endangered Species Act makes it a crime to "take" an endangered or threatened plant or animal species. An "endangered species" is one that is threatened with extinction, while a "threatened species" is one that is on the verge of becoming endangered. The act further defines *take* as "to harass, harm, pursue, hunt, shoot, wound, kill, trap, capture, or collect." The word *harm* is the most controversial as it relates to forest owners. The U.S. Fish and Wildlife Service, responsible for enforcing the ESA, defines harm as "significant habitat modification or degradation [that] actually kills or injures wildlife by significantly impairing essential behavioral patterns, including breeding, feeding, or sheltering." The ESA also extends to plants, but the responsibility of private owners to protect threatened and endangered plants is not nearly as imposing as the responsibility for wildlife, with one exception—if you have accepted federal funds for work being done on your land, plants have more protection under ESA than if no federal funds or permits are involved.

Because forest-management activities can result in "significant habitat modification," even if ESA does not apply, it is helpful to know if any ESA-listed species use your land. If forest management is expected to affect protected wildlife species, a landowner can still proceed with the project by completing a "habitat conservation plan" and applying for an "incidental take permit." The habitat conservation plan will help design your project so it has minimal effect on the protected species in question. It is an attempt to view your project from the context of a larger picture. The incidental take permit is legal protection in case the project results in a taking despite the habitat conservation plan. U.S. Fish and Wildlife Service biologists will help develop the plan, resulting in little or no extra cost for the project.

The United States has 960 species considered endangered or threatened, 431 animals and 529 plants. Most of the animals are aquatic species, and about half the plant species are forest dwellers. The threatened and endangered species of greatest significance to forest owners are birds, mainly in the West, in desert ecosystems, and in the Hawaiian Islands.

If you suspect that threatened or endangered plant or animal species are on your land, you should contact the U.S. Fish and Wildlife Service, Department of the Interior, Division of Endangered Species, 452 ARLSQ, Washington, DC 20240 (703-358-2171) to verify and to seek guidance on how to protect habitats during management activities. Some owners cringe at the thought of inviting the federal government to inspect their lands. After all, why would anyone knowingly risk having their hands tied behind their backs if a protected species is identified? Unfortunate as it may seem, society has a right to ensure the survival of protected species regardless of the land they occupy. If you own critical habitat, you have a legal and moral obligation to protect the species that live there if they are listed by the U.S. Fish and Wildlife Service as threatened or endangered. With a little extra planning and care—and good forest-management practices—it is easy both to use the forest and to protect important habitats.

The U.S. Supreme Court recently strengthened the rights of private individuals to challenge the Fish and Wildlife Service in its interpretations of the Endangered Species Act. Before its ruling, only environmental groups suing for more protection were allowed standing in such lawsuits. In effect, the Supreme Court has extended rights to challenge the act to any person who believes the Fish and Wildlife Service has overstepped its bounds. The result is expected to be a substantial increase in litigation involving development interests and environmental groups.

After more than twenty-six years, the Endangered Species Act is up for reauthorization, and provisions of the act are currently being debated in Congress. It is an often heated debate, which pits environmentalists against landowners and natural resource users, and it does not appear as if the differences will be resolved easily. Congressional committees are reviewing various provisions of the act, especially the process by which species are listed for protection, and deciding what constitutes a taking—both of an endangered species and its habitat and of an owner's real property rights. If you have opinions to offer, now is the time to contact your congressional delegation. The issues for reauthorization are so divisive, however, that it may take many years to redraft the Endangered Species Act.

## *The Clean Water Act*

Congress passed the Federal Water Pollution Control Act in 1948. In 1972, the act was amended to include pollution from "nonpoint sources," such as mud from logging roads, equipment lubricants, chemicals, and other pollutants. Most people think of water pollution as coming from pipes, also known as "point-sources." A nonpoint pollutant is not easily traced to the point of origin, but gets into the water as runoff, usually from farm and forest lands.

In 1977 the law was renamed the Clean Water Act (CWA), and it was reauthorized by Congress in 1987.

Administered by the Environmental Protection Agency (EPA), the CWA requires states to develop Best Management Practices

(BMPs) to protect water quality during timber harvesting. States are also required to develop programs to ensure BMPs are used, and to periodically report to EPA on the status of water quality. Although administered by EPA, enforcement is solely the responsibility of each state, and each state has approached its responsibilities under the CWA differently. A state's BMPs must consider "economic, institutional, and technical factors" so their design and installation is cost effective and reasonable (American Forest and Paper Association 1994). EPA's jurisdiction under the CWA extends to "all waters of the United States, including adjacent wetlands." This broad jurisdiction encompasses every acre of woodland in the country. Congress, in its wisdom, provides an exemption from the permit requirements for discharges that result from "normal farming, silviculture, and ranching activities such as plowing, seeding, cultivating, minor drainage, harvesting for the production of food, fiber and forest products, or upland soil and water conservation practices." In most states this means either (1) the BMPs promulgated at the state level must be followed, or (2) the BMPs are "highly recommended," but if a discharge occurs and the BMPs are not in place, the owner may be subject to fines. Because it is almost impossible to extract timber from most woodland sites without disturbing the soil sufficient to cause a discharge, it is in the owner's best interests to see to it that the BMPs are followed. Most of the BMPs (even though they differ slightly from state to state) are common sense, and their use makes timber extraction only marginally more expensive.

For more information on BMPs, contact your state forester. Also, many local extension foresters have publications that explain a landowner's obligations under the Clean Water Act.

## *Ecosystem Management*

Although a discussion of ecosystem management may seem out of place in a chapter dealing with the concept of private property and the public's interests in the rights of private owners, it is most germane to the subject. Managing whole ecosystems will redefine the nature and extent of property rights, and although the debate is still engaged, state and federal forest-management agencies—even forest-using industries—are changing their policies to protect and sustain important ecosystem structures and functions. Ecosystems do not recognize property boundaries, so forest owners need to know the effects ecosystem management practices may have on their bundle of rights.

Increased recognition of the interrelatedness of forest organisms and the potentially far-reaching and long-lasting effects of traditional uses of forests has spawned new ways of thinking in recent years. Although not fully embraced by all forestry professionals, some of whom contend that traditional uses correctly implemented can protect and improve forest ecosystems, the term *ecosystem management* implies three things. First, that primary importance is placed on maintaining the long-term forest health, not just for human use or even for future generations but for the well-being of plants and animals that also occupy and use forests. Humans are nearly co-equal users of the forest with the power to control disturbances to lessen the negative consequences and, where possible, duplicate circumstances that mimic natural processes. This is akin to saying that, although we might have the knowledge and power to dominate and fully manipulate woodlands solely for human use, we have a responsibility to protect and provide for all other organisms that rely on the forest (McEvoy 1995).

Second, ecosystem management implies that humans accept the responsibility to understand the complexities of forests and to err on the side of inaction rather than risk subtle but irreversible damage attendant upon proceeding in ignorance. Our management strategies, while attempting to mimic natural processes, must change as we learn. The days of stocking charts and universal prescriptions aimed solely at increasing the volume and commercial value of timber are nearly over. This is not to say forests will no longer supply timber. But it does mean that timber production is subordinate to the long-term health and sustainability of the forest. Fortunately, disturbance—sometimes catastrophic changes brought on by storms, insects, disease, and fire—is the rule in forests rather than the exception. Practicing ecosystem management means designing disturbances in a way that allows us to use the resource while protecting the integrity of the landscape.

Third, because ecosystems are recognized at many different levels—from small populations of fungi to whole landscapes—and do not follow geographic or political boundaries, practicing ecosystem management will inevitably modify our traditional ideas about real property rights (see figure 1.1). For example, some migratory songbirds that winter in tropical regions of the Caribbean and the Americas use the forest of North America during the spring and early summer for nesting. We know that relatively undisturbed, old-growth forests are important habitats for many of those species. Practicing ecosystem management to protect nesting opportunities for neotropical songbirds requires guidelines that will probably limit harvesting in certain forest types for a few weeks each year. To some people, that is a small price to pay to ensure the survival of those species. To others, it means an erosion of private property rights.

Because ecosystems vary in size and complexity, from the population dynamics of bacteria in a single forest soil to the global interactions of species—including humans—managing whole ecosystems will require an unprecedented level of cooperation among neighboring woodland owners, municipalities, states, and countries. Uncertainty about how this cooperation will occur, especially at local levels, is a primary source of concern for owners, managers, and users of forests. Successful implementation of ecosystem management worldwide will require drastic changes in forestry practice.

A short list of things we can expect to see, if ecosystem management is to be implemented on a global scale, includes the following:

- Catalogue species at global, continental, regional, and sublevels. Which species are indicators of ecosystem integrity and which species are destined to become extinct regardless of human activities?

- Identify and project human demands for wood and allocate demand to wood-growing areas of the world.

- Speed up the search for wood replacements. Some forest ecosystems that now produce a great deal of wood will produce considerably less than projected demand under an ecosystem management approach. Economical and environmentally friendly substitutes for wood will become essential.

- Clearly establish society's values regarding forests. What are we willing to pay for and what can we afford to give up?
- Redefine silvicultural practices in ways that mimic the natural processes of a particular landscape.
- Rethink timber extraction timing and methods on some sites.
- Increase our willingness to wait longer, to accept more costs for tending forests, and to pay more for wood products.

Some people contend that ecosystem management is a passing fad. Given a few years, a new administration, and substantially higher prices for wood products, we will be back to business as usual. Although it is true our resolve to protect whole ecosystems from excessive and exploitative human uses will wax and wane, most forest managers and academics agree that the lasting effect of ecosystem management will be a change in our thinking about forests, the resources they can provide, and the best ways to use and sustain them. Our concept of private real property will never be the same.

## Chapter 3

# Acquiring and Owning Forest Lands

When you acquire forest lands, you acquire whatever rights the seller or donor (in the case of a gift or inheritance) had in the land. Or you acquire only those rights he or she is willing to convey. For example, the owners may want to retain some of the rights for themselves (retaining mineral rights used to be very common in some parts of the country). The seller or donor may also want to protect against a future owner's attempt to exercise rights the current owner deems inappropriate, such as the right of development. Separating the bundle of rights is tricky business and requires the efforts of an attorney who specializes in property law. When an owner grants a separation of rights to another person or legal entity or, upon sale, retains some rights, this is known as an *easement*. Easements are discussed later in this chapter.

Before you acquire forest land, you want to be absolutely certain of the rights that go with it. For most land transactions, it is essential to have a local attorney verify the title to a tract of land and to identify any potential problems that may exist from past relationships among owners or from previous transactions. The actual title is nothing more than words that describe the extent of a person's rights. A deed is a document that describes the physical location of the property and the extent of an owner's rights. Whenever land changes hands, the buyer (or the buyer's attorney) must examine the tract's title to be sure the current owner has all the rights the buyer wants to purchase and that there are no undisclosed claims (or easements) on the current owner's title. In many states, before the transaction is completed, statutes require a lawyer's opinion on the title, as well as title insurance to protect against claims in the future. Real estate title is discussed further in the next section. Here, I want to discuss briefly the nature of a person's estate—the legal extent of an individual's interest in land—to help explain why the history of a tract's title is so important.

A title to a tract of land can be original or derivative. An original title is usually assumed to be vested in the government or the state and is acquired through discovery, conquest, or occupancy. In the eastern United States, the concept of original title is antiquated, but there are vast areas in the West and in Alaska where much of the land is under original title. A derivative title is vested in individuals and can be passed by sale, gift, or bequest. A contract to convey title, or an option to acquire title when certain conditions have been met, will create a title by devise, compelling the seller to pass his interests in the land at some future date as specified in the contract. (Title to real estate is discussed in more detail in the next section.)

The extent of ownership, or the conditions under which one has a claim of rights and the duration over which those rights belong to a person, is known as the *estate* of the owner. There are two primary forms of estate, defined by the presence or absence of imposed time limits:

1. A *freehold estate* has no fixed time limits on the current owner's rights, and the rights can be passed without time limits to another person. Most land is owned as a freehold estate. It represents the greatest extent of rights available to a property owner and it is the least restrictive form of ownership.

2. A *less than freehold estate* is a situation where there is a specified time period during which the person who holds the estate retains rights. For example, a lease creates a less than freehold estate.

Aside from the obvious differences between these two forms of estate, a freehold estate usually encompasses the full bundle of rights, and a less than freehold estate encompasses something less than the full bundle of rights. A long-term lease to tap conifers for the production of naval stores is a less than freehold estate. The person holding the lease (the lessee) has the right of access and the right to gather sap, and no other rights. Usually this type of estate has an expiration date after which the lessee must relinquish all rights. Another example of a less than freehold estate is a property from which mineral rights have been separated.

A property can encompass more than one estate. An owner of a freehold estate can create a less than freehold estate vested in another person, as in the example above involving a lease for naval stores production.

Under the concept of freehold and less than freehold estate, there are seven different forms of estate that define the extent of an estate holder's rights, arranged below from the most inclusive and extensive bundle of rights to more restrictive estates, where the person has rights but not necessarily by virtue of ownership.

1. An *estate in fee simple* (or just *fee* or *fee simple absolute*) encompasses the most extensive bundle of rights available to an owner. Most tracts of land in the United States are held as estates in fee simple. Unless the current owner places restrictions on the title, the buyer (the beneficiary or heir) can assume he or she is acquiring a freehold estate in fee simple.

2. A *life estate* can be an estate in fee simple, but the current owner's rights expire when he or she dies. The essential point is that a life estate cannot be passed on to heirs. A life estate is a less than freehold estate in fee simple. Or, it can be a less than freehold estate in less than fee simple if the current owner has some rights but not all.

3. An *estate in remainder* encompasses all of the rights that were available to a decedent before he passed away. These may include all the rights of a fee estate, partial rights, or no rights at all if someone else can prove a better claim. An estate in remainder may be freehold or less than freehold.

4. An *estate in reversion* means the current owner's rights have expired and the estate "reverts" to someone else. For example, when a long-term lease expires, the estate created by the lease reverts to the person who granted the lease.

5. An *estate from year to year* (or month to month) is a less than freehold estate that usually exists for a year or less. If an owner (the lessor) were to lease hunting rights one year at a time, it would create (for the lessee) an estate from year to year. Unless a termination date is specified, the lessor must notify the lessee that the lease has ended.

6. An *estate at will* is a less than freehold estate that can be terminated at any time by the person granting the rights or by the person holding the rights. For instance, an agreement granting a neighbor the right to pick strawberries on your land may be an estate at will for the neighbor, because you can revoke those rights at any time. An estate from year to year becomes an estate at will if the lessor is silent as to the ending of the lease and no termination date was specified.

7. An *estate at sufferance* is a less than freehold estate where the conditions by which the rights were originally granted expire, but the person granted the rights continues to use them "at the sufferance of the owner." For example, a long-term lease expires and neither the lessor nor the lessee bothers to renew it. The lessee continues to pay rent and the lessor, by his or her actions, accepts the conditions.

The law also recognizes two different types of conditional estates. An estate in fee simple "determinable" is a conditional estate that exists for a duration as long as certain conditions are maintained. The conditions are usually prefaced with the words *so long as*, *while*, or *during the period that*. If the conditions change, or the period expires, the estate reverts to the grantor or his heirs. An example might be a gift of forest land, as long as it is used for educational purposes. The land can never be sold for any other uses, and the estate is also conditioned on the forest being used for educational purposes. A fee simple determinable estate ends automatically when the property is no longer used for its intended purpose.

Another conditional estate is an estate in fee simple "on condition subsequent." Words like *if*, *provided that*, or *on the condition that* indicate the allowed conditional uses of property. *Condition subsequent* differs from *fee simple determinable* in that the words make no reference to duration. It also differs in another major way. An estate in fee simple determinable reverts to the original grantor automatically when the conditions are not met or a specified period ends; an estate in fee simple on condition subsequent must be reclaimed by the original grantor or the grantor's heirs for the rights of the current owner to end.

## Title to Real Estate

A title as it relates to real estate is an abstract concept. You cannot hold a title to real estate in your hands, yet it is very real and significant because it expresses a person's legal rights to land—rights as evidenced by a deed, a survey, or some other legal documentation, such as a contract or a bill of sale. A person acquires forest land by obtaining a legal title to it, one that no other person can challenge. The process of passing title is called "alienation." A current owner must alienate the title to pass it to a new owner. The title to property defines the extent of a person's estate.

The laws surrounding the private ownership of real estate are exhaustive and often confusing, and they vary from state to state. Law students devote an enormous portion of their time to the study of real property law, but not all lawyers make a practice of handling title issues. Anytime you anticipate changing the title to your land, or you are presented with a claim against a title you hold, seek the advice of a competent local attorney. Among attorneys, the finesse with which a title is passed is the legacy by which their future peers will judge them.

A tract's title is the equivalent of an animal's bloodlines to a breeder. Both the purchaser of real estate and the person shopping for particular traits in an animal look to the past to ensure they are getting exactly what they are paying for and there are no problems that may crop up in the future (due to error conveying title in the past or—for a breeder—a mismatch of bloodlines). An error in a tract's title, or an ambiguity caused by marriage, bequest or gift, or a title acquired by fraudulent means will create a "clouded title." A clouded title must be cleared up before a new owner acquires it. If it is not, the new owner faces the possibility of having the claim challenged by an earlier, legally valid claim. The courts will recognize the earliest legal claim on a tract's title, which may be from the grandchildren of the person who incorrectly alienated title in the past—the person you purchased the land from. For this reason, it is essential to have an attorney or title company search the title of the tract you are acquiring and to suggest legal methods to clear up any clouds on the title. Even in states where title insurance is not required, it is money well spent to protect against losses from title claims of others. Title insurance is a one-time fee paid at the time title changes hands.

As discussed earlier, most real estate titles are derivative, which means their history over the past few generations is traced by the legal alienation of rights by former owners. These rights are vested in a new owner, who has then "acquired" title. Alienation is usually voluntary through a legal conveyance of deed, such as when property is sold or willed. Sometimes title is acquired by involuntary alienation, however, without the current owner's consent. Involuntary alienation results from actions such as the following:

- *Eminent domain.* A (local, state, or federal) government's right to take land for the public good.
- *Escheat.* The state's right to claim the land of a decedent who has died without a will and no legal heirs.
- *Foreclosure.* A court-ordered process where property is sold to repay a debt on which the property holder has defaulted.
- *Adverse possession.* Any action where an owner's claim to property is passed to another against his or her will by action of a public entity, through eminent domain, by action of a court judgment, or through some other process whereby another individual stakes a claim against an owner's property.

An adverse possession will create an original title vested in the person who claims ownership by right of occupancy and use of another's property for a statutorily defined period of time (which varies from state to state, but seven years is common). According to the Real Estate Licensing Manual (1996), acquisition of title by adverse possession "must be by the exclusive, notorious, hostile, continuous, physical, and open use of the property contrary to the best interests of the true owner." Evidence of an adverse possession usually must include the requirement that the person making the claim has been responsible for paying property taxes on the tract in question. And the use must be contrary to the interests of the owner. For instance, a caretaker living on the property cannot make a claim by reason of adverse possession unless it is clear the owner has relinquished his or her rights. A squatter living on your land has no rights to acquire title by adverse possession unless all of the above conditions are met. Adverse possession applies only to private lands, and it is the only method available—short of future conquests—to create an original real estate title.

## Deeds

A deed is a legal document used to convey a person's interests in land. A deed must be a written legal description of the property so the location of the tract and the rights that go with it are clear to everyone (including future generations). The current owner, or grantor, must have a legal capacity to convey clear title, and the receiver, or grantee, can be any legal entity (person or company)—even minors have a legal capacity to acquire title. The deed, as a form of contract, must also have the following:

- A clause that actually conveys title from grantor to grantee.
- A clause demonstrating that consideration has been paid by the grantee for the rights the grantor is about to convey. The consideration clause is usually a nominal statement, such as "For the sum of $1.00 and other valuable consideration . . ."
- The legally witnessed and publicly acknowledged (witnessed by a notary) signature of the grantor. Unlike other types of contracts, it is not necessary for the grantee to sign the deed.

After the deed has been executed, it must be delivered to and unconditionally accepted by the grantee for it to be valid. To effect a legally valid deed, the delivery must also be unconditional and voluntary. Delivery cannot be conditional on future events or circumstances, and it must be clear that it is the grantor's intent to immediately convey title to the grantee. Usually, if the grantor's signature has been notarized, it is assumed delivery is voluntary and immediate and that the grantee has accepted the deed without conditions. Once the deed is delivered, it is usually recorded in the town or county where the property is located. Even if the deed is lost before it is recorded, however, the grantee still holds title to the property (at least with respect to the grantor's conveyance, or passing, of title to the grantee).

The purpose of recording the deed is to publicly acknowledge the grantor's conveyance and to establish the grantee's claim to the property, since priority is given to the earliest legally valid claims to title. An unrecorded deed is invalid with respect to earlier recorded claims by other parties against the same property. For example, a grantor conveys title to forest land while a timber sale is in progress. On the day the deed is delivered, the grantee tries to stop the sale. When the grantee goes to record the deed he or she discovers the timber buyer recorded the timber sale contract months ago, creating a claim (a title by devise) that precedes the grantee's. Who owns the timber? The timber buyer—at least to the extent it is defined in the recorded contract.

A title search preceding the transaction would have discovered the contract. If there was no record of the contract, who legally owns the timber after the grantee has recorded the deed? The grantee! The timber buyer must seek recourse from the grantor and cease harvesting once he has been given legal notice of the grantee's claim. A properly recorded deed will take precedence over an earlier deed that has been delivered but not recorded. An unrecorded deed is invalid against subsequent claims that have been properly recorded. Since town and county records are open to the public, the only way to keep track of and discover valid claims to property is to record with the municipality legal documents that create claims.

Deeds and other legal instruments (such as a long-term timber sale contract in the example above) are recorded chronologically by reference to volume and page of the town or county's property records. An alphabetical card file (or electronic database in more sophisticated and better-funded municipalities) cross-indexed with the names of grantors and grantees provides the person searching a title with the first reference to a property in question (figure 3.1). The system varies from place to place, but the ability to cross-reference names and claims is essential. A title search of the grantor in the example above will yield a notation (or card) in the timber buyer's name (being the grantee). The notation will give the volume and page number where the timber sale contract is recorded. Any other claims against the grantor's title will be located in the card file in sequential order, beginning with the first claim (which takes precedence over all subsequent claims). The grantor's original title will be cross-referenced to the name of the person from whom he or she acquired title and the volume and page where that title is located. In this way an abstract of title can be methodi-

*Figure 3.1. Deeds, surveys, contracts, and other real property documents are filed with the town or county and can be located by a cross-reference of the names of grantors and grantees. Some communities use a card file index or database to locate the volume and page of the documents in question.*

cally searched back to a grantor so far removed from the current grantor as to ensure no legal claims to the tract's title.

In most states, statute defines the minimum period over which a title must be searched. Unless there are ambiguities because of boundaries or easements granted in the past, the search usually extends over the past thirty to sixty years—about two generations.

## Types of Deeds

There are many different types of deeds, defined by the purpose for which title is being conveyed. The four most common deeds involving forest land transactions are the following:

1. *Warranty Deed.* The grantor accepts liability for defending the claim he or she conveys to the grantee. It is the best deed a grantee can accept because it is, in effect, a guarantee that the title is good and clear. A warranty deed will include an affidavit of the grantor claiming that the title is clear and marketable. The grantor will also give assurances to the grantee that there are no encumbrances on the title other than those listed in the deed. Finally, it is customary that the grantor will agree to cooperate in the future if it is necessary to execute other documents to clear up any clouds on the title, a process known as "perfecting title."
2. *Mortgage Deed.* The grantor (mortgage purchaser) acknowledges the claim of a lender (as grantee) and pledges to conserve the value of the property and to accept financial liability to the lender—as secured by the property—at least to the extent of the mortgage balance. The mortgage deed (not used in all states) is a means of putting the lender's interests first if the property is sold or claimed by adverse possession. Banks, however, are often reluctant to loan money for the purchase of forest land unless the fair market value of the land (usually for development) greatly exceeds the loan amount and the borrower has other sources of income to repay the loan. In some states, a mortgage is treated as a lien against the deed, not an ownership interest in land.
3. *Quitclaim Deed.* The grantor agrees to convey only the interests he or she might have in a property. Because the grantor's claims may be questionable, he or she does not warrant the title in any

way. The quitclaim deed, as the name implies, is often used to eliminate confusion about claims resulting from marriage, divorce, name changes, or other circumstances—such as a poorly marked boundary—where there is (or might be in the future) ambiguity about who owns what.

4. *Gift Deed.* Similar to a quitclaim deed and used to convey title by gift. A gift deed can be invalidated, however, if the grantor is making the gift to avoid creditors, including the federal and state government if the gift is made in anticipation of death and solely for the purpose of avoiding estate tax liability. (Estate taxation and tax avoidance strategies are discussed in chapters 8 and 9.)

Other, less common, deeds include a trustee's deed, an executor's deed, a tax deed, and a sheriff's deed. In each instance, title is being alienated by a legally appointed grantor who has been given the right to convey title.

## *Easements*

An easement grants rights to land without actually owning the land. In most circumstances, the grant is long-term or permanent. It is the most common method of separating the bundle of rights, especially for the purposes of protecting land from development. The holder of an easement is called the "grantee," and the person on whose land the easement applies is the "grantor."

The grantor of an easement conveys rights to a grantee usually for very specific purposes, such as for a driveway or for drainage purposes. An easement, though, is usually not temporary. It stays with the land and is passed on to future owners. The grantee of an easement, if it happens to be an abutter, should have a reference in the deed regarding the terms of the easement. When the grantee's property is adjacent to the parcel with the easement, it is called an "easement appurtenant." In this case, the grantee's property is the "dominant estate" and the property of the easement's grantor is called the "servient estate."

When someone has the right to use land that is not adjacent, it is called an "easement in gross." An easement granted to a municipality for sewer lines that cross your property or to a power company for high-tension lines are examples of an easement in gross. Another example of an easement in gross is when development rights are sold or given to another legal entity, such as a land trust. (Easements as they apply to land trusts are discussed in more detail in chapter 9.) When an easement is given or sold to a land trust for the purpose of forever protecting land from development, it is also known as a "transfer of development rights," or TDR.

The conditions of easements in a property title are also known as "covenants." A covenant is a promise—in the case of an easement, a promise to honor and protect the terms as originally set forth. An owner who acquires land with covenants accepts the promises of a former owner and agrees to honor the terms as though the promises were his own. A deed may also include other covenants made by the grantor, such as a covenant that the title is free from encumbrances. When a covenant creates restrictions on the titleholder, however, it is known as a "deed restriction." Any condition referenced in the deed that limits in any way the titleholder's use and enjoyment of the full bundle of rights is a deed restriction. For example, a requirement that a deed holder maintain the color scheme of a house because of its historical significance is an example of a deed restriction imposed by a municipality. Another example of an easement is a sewer line or power line crossing your property.

An easement for conservation purposes is the most common method of relinquishing development rights on farm and forest lands. When the grantee of the easement is an IRS-qualified, nonprofit organization dedicated to holding land for conservation purposes, the original grantor may take advantage of Internal Revenue Service rules that can result in substantial tax savings over several years. The tax advantages, however, are available only when the easement is made for conservation purposes and in perpetuity. This type of restriction is often used to limit development on farm and forest lands. It does not restrict locally acceptable forest-management practices, and aside from the conservation easement, the land can be sold, willed, or given as an estate in fee simple absolute. (The tax implications of easements for conservation purposes are discussed in chapter 9.)

## Methods of Holding Property

There are five principal methods of holding property, but not all five are recognized in all states. The differences among methods have to do with the number of different people (or legal entities) involved and their relationship to one another (figure 3.2).

Property held *in severalty* means there is only one legal entity—an individual or a corporation.

*Tenants in common* involves ownership by two or more people, each of whom is entitled to an undivided possession of rights even though the total ownership involves the sum of separate interests. Without specific mention of proportional interests, it is assumed the interests are equal. Also, it is usually assumed the interests of the owners apply to the entire property even though

*Figure 3.2. Property can be held many different ways. Knowing the best method to hold assets with others is an important consideration for an owner of forest land.*

separate interests can be defined. The tenant in common can sell or lease his or her interests in the land to another, and can will his or her interest to heirs. Often a decedent's estate will pass to heirs, who become tenants in common. An heir wishing to sell interests to the other heirs would usually do so with a quitclaim deed.

A noncorporate *partnership* is most apt to be owned by tenants in common. A tenant can sell his or her interests at any time unless provisions in the partnership agreement prescribe how the tenant in common must sell.

*Joint tenants*, or *joint tenants with rights of survivorship* (JTRS), is an ownership by more than one person where possession of the property and interests are shared equally through one title of a freehold estate (no time limits). This is a common way for married couples to own property (in many states where common law is recognized), although there can be more than two joint tenants. The separate interests of the parties cannot be willed, and when there are survivors, the property of a joint tenancy is not probated. When one tenant dies, his interests pass automatically to the other joint tenants. In some states, a joint tenant can sell his or her share of the interests without consent of the other parties, but the person who acquires title does so as a tenant in common with the other joint tenants. If the shared interests of joint tenants are to be sold, the parties must act together. A joint tenant with rights of survivorship usually cannot sell his or her interest to a third party. Without an agreement, a court order is necessary to split up the rights of joint tenants.

A *tenancy by the entirety*, not recognized in all states, is only available to married couples. It is as though the couple were one person, thus there is no such thing as equal shares, and one party cannot sell (or will) interests in the property without full consent and cooperation of the other. Tenancy by the entirety implies rights of survivorship, and the surviving spouse will hold title in severalty. In the event of divorce where the property must be divided for settlement, the title can be vested—by stipulation of the parties or by action of the court—as joint tenants or tenants in common.

In some states (Arizona, California, Idaho, Louisiana, Nevada, New Mexico, Texas, Washington, and Wisconsin), property acquired jointly by spouses during the marriage is treated as *com-*

*munity property.* Any property brought to the marriage, or acquired by gift or inheritance after the marriage, is separate property (held in severalty by one spouse or the other). The property must be maintained by separate funds and not by commingled funds of the spouses. Community property is assumed to be shared equally and can be willed to other parties, who become tenants in common with the surviving spouse.

Contrary to community property, in some states "marital rights" are recognized. These are rights (usually a percentage of the entire estate) that extend to a spouse regardless of the independent actions of the other spouse. For instance, when a husband dies with property held in severalty, the wife has dower rights that may supersede a bequest of the property to someone other than the wife. *Curtesy* is the same rights extended to the husband. If either spouse predeceases the other, all marital rights die with them.

## Owning Land with Other People

Most private nonindustrial forest land is owned by married couples. In many instances, however, individuals or couples pool resources and purchase land with other individuals. They do this by forming either a corporation or a partnership.

## Corporations

A *corporation* is an ownership in severalty. Individuals incorporate the assets of a forest holding to create a single legal entity. There are some advantages to holding forest assets as a corporation, especially if the holdings are extensive and several people share an interest in the land.

The six criteria used by the Internal Revenue Service to define a corporate entity are as follows:

1. An association of individuals that have a stake in the corporation, also known as *shareholders.*
2. A profit motive.
3. A centralized and collaborative management structure, usually involving a board of directors who make day-to-day decisions for the corporation.

4. Longevity that exceeds the lifespan of the shareholders. In a sense, a corporation has its own "life."
5. The ability of shareholders to transfer their interests to others without restrictions by the corporation.
6. Limited liability for the shareholders due to the actions of the corporation.

The biggest advantages of a corporation are that shareholders can come and go, and although they share in the profits (and losses), they are protected from liability resulting from the actions of the corporation. Incorporating forest holdings among family members is a strategy to lower the value of a person's estate to avoid estate taxes (see chapter 9). Of the two primary profit-oriented methods of incorporating recognized by the IRS, Subchapter S is most commonly used by individuals and families. In a Subchapter S corporation, the assets are "closely held" and there is no public offering of shares to individuals other than those defined in the corporate charter. The major advantage of a "Sub S" is that profits or losses go directly to the shareholders' tax returns. Thus, profits are taxed only once and losses can offset income from other sources. On the other hand, profits earned but reinvested are still taxable to shareholders, who then must come up with the cash to pay taxes due. Setting up a Subchapter S corporation is fairly easy, but it requires the advice, guidance, and skill of an attorney with this type of experience. People interested in Sub S corporations should review the small-business reference section at their local bookstore for more detailed information on their advantages and disadvantages of incorporation.

## Partnerships

A *partnership* is a noncorporate association of two or more people, usually as tenants in common, who are personally liable for the actions of the partnership. There are two types of partnership: general and limited. In a *general partnership*, the individuals are equally (or proportionally) involved in the day-to-day operation, sharing in the profits and losses and being jointly and severally liable for the acts of the partnership. In a *limited partnership,* there are two kinds of partners: the general partner(s), who controls the operation, and the limited partner(s), who have no management authority and only limited liability for debts and other claims against the partnership (Shumate 1995). A limited partner is often the source of capital for an operation, sharing in the profits and losses (with the advantage of writing off the losses against other sources of income at tax time).

A partnership of two married couples creates a tenants in common ownership with respect to the couples, even though the husband and wife are joint tenants with respect to their combined share of the partnership.

Family partnerships are a common way for parents to maintain control of a forest property (as general partners), while dispersing the assets to children, who become limited partners (creating a family partnership as an estate-planning tool is discussed in more detail in chapter 9). In recent years, the IRS has given much closer scrutiny to family partnerships set up ostensibly for the purposes of avoiding taxes.

## Purchasing Woodlands

There are three ways to acquire forest land: by purchase, by gift, or by inheritance. Most of the 9.9 million private forest holdings in the United States have been acquired by purchase, but increasingly, lands are being passed through inheritance. Because the average age of forest owners is increasing, and owners are holding their lands longer, the trend toward inheritance is expected to continue. Unless heirs *want* the land, however, an inheritance may result in a quick sale to help settle the estate. Gifts of forest land are fairly rare, but aging forest owners will often consider the possibilities of giving land to children to help lower the estate value. (Leaving forest land to heirs and making gifts of land to family members are discussed as estate-planning strategies in chapter 9.)

There are many forest land buyers each year. Most do so primarily because the land is associated with a house they want to buy or with a parcel on which to build. Many buyers, though, look to acquire forest land for the sake of long-term management of forest resources. Acquiring forest lands by purchase requires good planning and careful evaluation of many details. What follows is a brief discussion of what to look for in forest land, ideas on how to pay for it, and important questions to ask the seller.

The best buy in forest land (assuming timber production is one of the buyer's objectives) is a relatively young forest that will reach maturity within the next ten to fifteen years. Why? Because the timber has not reached merchantable size and thus its contribution to the appraised value of the land is negligible (figure 3.3). If you can wait, the forest inventory will go from almost nothing to substantial merchantable volumes within a relatively short period. It is also easier for a trained eye to see productive potential in young stands. Older, larger stands can mask poor productivity with larger trunks. A buyer with an eye to timber potential will look for tall, pole-size trees supporting healthy crowns in easily accessible stands. Some buyers are tempted to buy land with saleable timber, hoping to sell the timber and pay back the bank (or the parent). This may seem like an excellent strategy, but it is not. After all, who wants to live with heavily cutover lands for the twenty to forty years or more that it will take for the timber to grow back? A local consulting forester can be an invaluable source of assistance in locating a tract of forest that is a good buy.

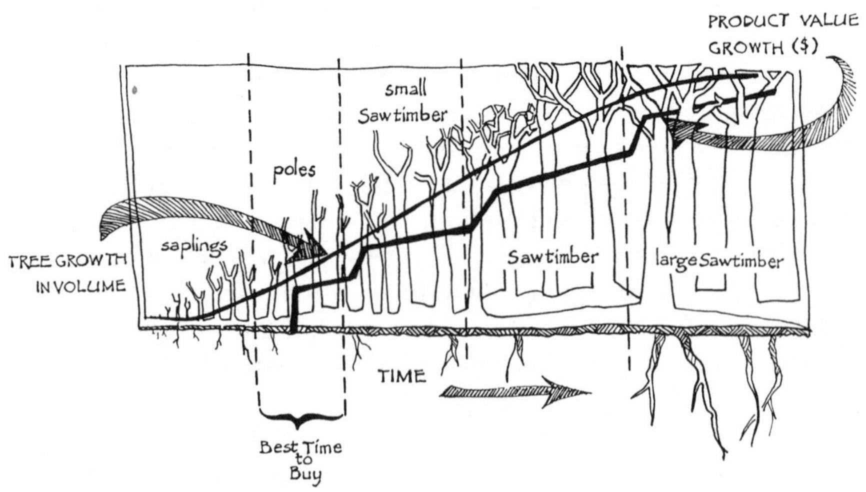


*Figure 3.3. Forest volume growth and value growth change at different rates as stands get older. The best time to buy forest land is just before trees reach merchantable size.*

Another important consideration is access. Does the land front directly on roads, or is access gained via easements through surrounding lands? How good are the roads, and are there any access restrictions? Are there any rights-of-way on the land, and what are the prospects for future public claims for access? What about the extent and condition of roads on the property? Are they well designed and maintained? Are there any erosion problems? Do streams run through the property? What about stream crossings and the condition of bridges and culverts? Is there a source of sand and gravel on the property? Roads need periodic upkeep, and a local free source of gravel will seem like a gold mine.

Some other important questions have to do with boundaries. Are the property boundaries clearly established and marked? Are they accepted and recognized by neighbors or are disputes brewing?

Something else to consider: Is there a good, accessible building site on the property, if you should decide to build a house? Will the soil allow a conventional septic system, or will it require special (and expensive) designs? Is good-quality water available at a reasonable well depth near the house site? Are any zoning restrictions in place that will limit your ability to build? Are there any sources of contamination in the aquifer you are apt to tap into? Generally, look uphill for sources of pollutants—it may be necessary to obtain the opinion of a geologist.

Also, before you make an offer to purchase the land, be sure no toxic dump sites are the property. Look near barns and toolsheds for areas where no vegetation grows, or where the vegetation is sparse and distinctly different from surrounding areas. Dig around and ask questions, because when you buy property you also buy any dumps that are on or in it and any future liabilities arising from them. The title search should include an opinion about former activities that might have produced hazardous wastes. Before the closing, arrange to have all drums, fluid containers, and other non-inert materials removed from the site, and have the seller sign an affidavit stating there are no toxic dumps on the property.

How are you going to pay for the land? Most banks are reluctant to loan money for forest land unless it offers an express and obvious profit-oriented business opportunity. The exception to this is when the land is to be used for a primary residence; otherwise, expect to put up collateral for a bank mortgage and to pay a higher interest rate over a shorter loan period than for a primary residence. Sellers are often willing to help finance the purchase of their forest land, usually with a *land contract* or a *contract for deed*. Both are creative methods of owner financing, where the seller retains title to the property but agrees to convey title to the buyer on or before a specified date, or when certain conditions have been met. The land contract and contract for deed, however, create a title by devise for the buyer, because they document the buyer's first interest in the tract's title when the contract terms are fulfilled. Financing in this type of arrangement is set up to be affordable for the buyer, but with the stipulation that the entire balance due is paid before the end of the loan amortization period. Known as a *balloon note*, the loan is amortized over a long period—to make monthly payments affordable. But the note comes due, or balloons, on a date well before the end of the loan amortization period. The buyer must locate another source of financing at that time or make new arrangements with the seller. For example: A couple agrees to purchase 160 acres of forest land from a seller who is willing to finance the purchase with a note that balloons in five years. The purchase price is $80,000, and the couple puts $10,000 down and finances the balance, $70,000, with the seller at 7.5 percent amortized over fifty years. Monthly payments are $448.16, but the sixtieth payment is $69,242—the balance of principal due. The couple must refinance or otherwise pay the balance at the end of five years.

Finally, there is the question of how much to pay for woodlands. Your offer should be based on the going price for forest land in the region and on how you intend to use the land. A local consulting forester can help suggest a fair price. The more land you buy, generally the lower the price per acre. The land probably will be priced as though it were to be developed, which is substantially more expensive than if the land is to be used strictly for forest values. If you intend to maintain the land as forest, you may be able to persuade the seller to sell (or give) the development rights to a local land trust, or you may agree to pay fair market value and sell or give the rights after you own the land. (The sale or donation of development rights is discussed in chapter 9.)

# Chapter 4

# Surveys and Boundaries

The number of people who own forests and do not know the boundaries of their land is surprising. This situation is more common in states of the original thirteen colonies, where the method of survey is based on landscape features, vegetation (which can change over the course of a single generation), and surveyor-placed monuments. Known as metes and bounds, and described in greater detail below, the primary problem with this system is that every parcel has its own reference, which may or may not agree with the references of surrounding tracts. It is easy to see how confusion about property lines can arise with metes and bounds. In fact, boundary ambiguities are one of the primary causes of title imperfections in the region.

West of the Allegheny Mountains and in areas of the country that were not widely settled before the Revolutionary War, lands are surveyed based on a system enacted by the U.S. Congress in 1785. This system relies on "benchmarks" (references) common to all properties and defines parcels that are exclusively rectangular. Appropriately named the rectangular system, it was adopted by Congress because once a point defined by the intersection of north–south, east–west lines is established (the benchmark), it is as though an imaginary grid were laid over the land defining every possible parcel, even though the corners of any particular parcel are not marked. Such a system enabled Congress to quickly lay claim to vast tracts of land and to grant claims to Revolutionary War veterans. Because the orientation of tract lines is always north–south, east–west, and corners are approximately 90 degrees, locating and marking a parcel is fairly easy. The rectangular system is described is greater detail below.

The boundary of a property is more than just the perimeter. It also serves as a dividing line between parcels. People who share a boundary are known as abutters. If they agree on the boundary, there are no problems. If they disagree, a survey by a licensed surveyor is the only way to resolve the dispute. A survey, besides establishing boundaries on a map, also provides a legal and (you hope) defensible description of a parcel and its location relative to other ownerships. A copy of the survey is recorded in the municipality as part of the description of the title to a parcel. The earliest recorded, error-free survey takes precedence over all subsequent surveys. In the eastern states, a surveyor of a difficult parcel is apt to spend almost as much time title searching in the public records as reckoning lines and corners in the field.

## Metes and Bounds

The colonists needed a quick way to lay claim to land. Accuracy was not important as long as earlier claims were respected. A valid claim needed obvious markings in the field and a description of the marks recorded in the public records. Landmarks and natural boundaries, such as ridge lines, rivers, and lakes, were often used to define a property boundary. Where there were no prominent landscape features to note, a settler created his own boundary marks using blazes on tree trunks, or by making rock piles and stone walls. Often the distances between points were paced or unmeasured and the bearing of lines only vague references to a feature off in a particular direction. Among those early surveyors who used a compass, some made measurements based on true north, others based on magnetic north—an automatic error of up to 12 degrees in some parts of the country. Magnetic north is also known to wander from one decade to the next. Modern instruments correct for the declination of true north from magnetic north.

Interpreting a metes and bounds survey (the term comes from *metes* for measures of distances and bounds for directions) is similar to following the directions of a treasure hunt (figure 4.1). The

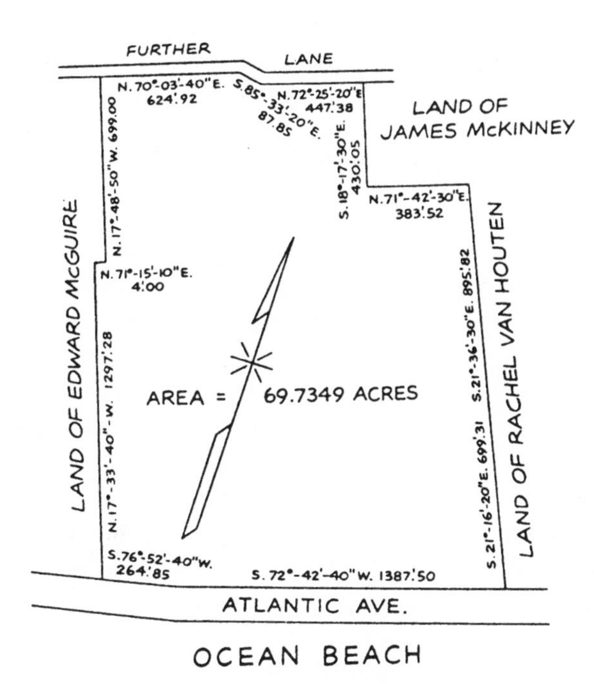


*Figure 4.1. A survey map with a metes and bounds description.*

inaccuracies, inconsistencies, and ambiguities of the past have become the bane of present-day surveyors.

A modern metes and bounds survey has the following features: a distinct and easily located starting point, definite corners or places where the line changes direction, accurate distances between corners or points, and bearings to the next major point or corner back to where the survey closes at the beginning. A surveyor will also estimate the total acreage within the boundaries and provide an estimate of the error of closure—where the beginning and the ending are supposed to meet exactly. Precisely measured angles and distances applied to legally valid and reasonably accurate earlier surveys are the methods by which surveyors today create clear, obvious, and irrefutably accurate surveys. Even modern surveys, though, are not without error. The difficulties of measuring distances and angles over uneven terrain coupled with small, seemingly insignificant measurement errors over the entire course of the survey can make the beginning and ending points impossible to close. The surveyor's goal is to create as small an error of closure as possible. Generally, the more accurate a survey need be, the smaller the error of closure and the more expensive the survey.

### *Establishing Boundaries in the Field*

A surveyor's job is to verify accuracy of the current title and to correct any ambiguities by researching earlier titles to the same parcel and to abutting parcels. The surveyor will also locate and mark corners in the field and provide the owner with a certified map that shows the parcel's survey and boundaries relative to surrounding parcels. Marking boundary lines in the field is not usually the surveyor's job unless there is a prior arrangement with the owner. Corners are marked by iron pins driven into the ground (or by a hole drilled into rock ledge, with or without a pin) so as to make it virtually impossible to move or alter. Even so, it is common practice to hide a corner pin and reference its actual location by triangulating from two or more surrounding landmarks (figure 4.2). This pre-

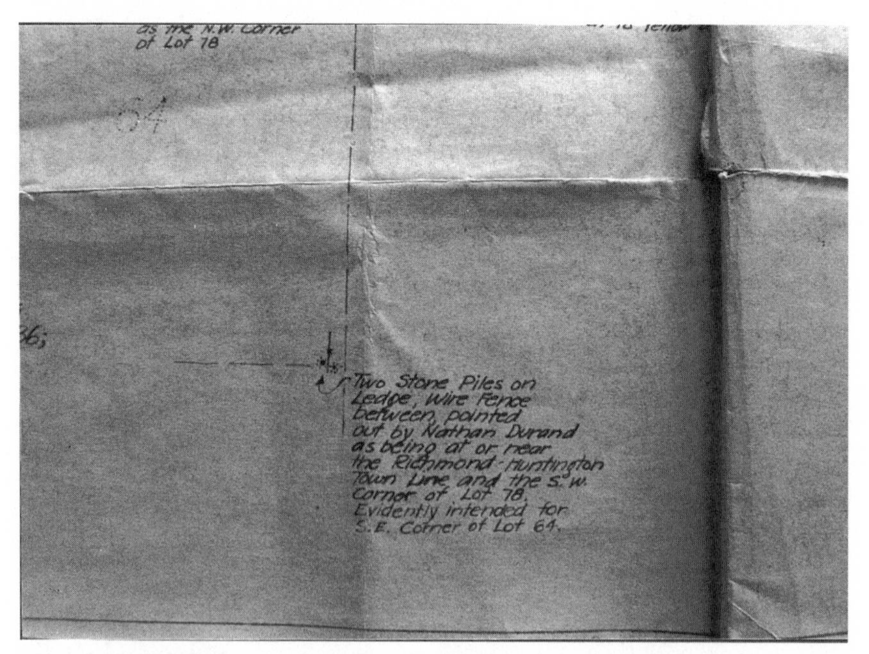


*Figure 4.2. Corners in a metes and bounds survey map are usually referenced to easily located reference points, such as large trees, boulders, ridge lines, or other prominent landmarks.*

vents tampering, which is illegal. State law in most places makes it unlawful for anyone other than a licensed surveyor to tamper with boundary markers.

When the boundaries of a woodlands survey are marked, the line between two corners is blazed. A surveyor makes a blaze by removing a small patch of bark from trees on or near the property line (figure 4.3). The top and bottom of the blaze are angled, and the exposed bare wood is scored along the same angle to make it as obvious as possible—ten or twenty years in the future—that the mark was intentionally caused by human effort. A blaze on a young stem will eventually close as the tree grows, but a trained eye looking for a boundary can usually pick out even the most overgrown and indistinct blaze (figure 4.4).

A boundary blaze faces the actual boundary line except when the tree is on the line (figure 4.5). A ring of blazes marks a corner, and when those trees are noted in the survey, they are referred to as witness trees (figure 4.6).

*Figure 4.3. Trees are blazed at about chest height with a blazing axe in a way that shows the marks were clearly caused by tools. After blazing, the marks are usually painted.*

*Figure 4.4. To a trained eye, even an overgrown blaze is fairly easy to discern.*

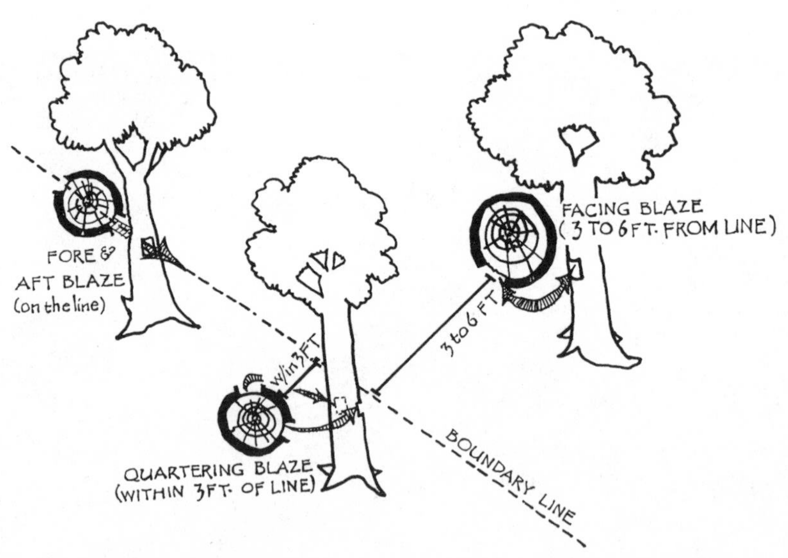


*Figure 4.5. When a boundary line is blazed, the location of blazes on trees is an indicator of where the actual boundary line is located. Blazes fore and aft mark a tree that is on the line; quartering blazes are located on trees that are within 3 feet of the boundary; facing blazes mark trees that are more than 3 but less than 6 feet from the boundary, and the blaze faces the line.*

*Figure 4.6. In modern surveys, corners are usually marked with iron pins (often abbreviated as "IP" on the survey map). The corner pin is referenced by multiple facing blazes on at least three trees. If the pin is disturbed, the corner can be easily relocated by triangulating from the witness trees.*

To improve visibility, blazes are often painted, though this is usually the work of the landowner or someone other than the surveyor. In most states, only a licensed surveyor can blaze a boundary or refresh an existing blaze, but a landowner or his or her agent can periodically paint blazes to improve visibility. Repainting blazes needs to be done at least every ten years, more frequently in younger stands. Use a good-quality, oil-based paint in red or some other bright color. Apply paint liberally and work it into bark fissures around the blaze. Plan to use a gallon of paint per mile of boundary and to dispose of the clothes you are wearing, and the brush, when done. The purpose of painting blazes is to make them obvious and easy to spot and doing so is a messy job.

Check with your State Forester or municipal official before refreshing blazes if you are not sure about the laws relating to property boundaries in your state.

Boundary trees are sacrosanct. Once a tree has been blazed, it is a boundary tree for as long as the tree stands or until the line is remonumented by a surveyor. In states where metes and bounds is used, a blazed tree is a property monument. It can be altered only by the actions of a licensed surveyor. Boundary trees should never be damaged or harvested during a timber sale.

When the survey is completed, be sure it is appropriately certified by the surveyor and that a copy is properly recorded in the town records. The survey is an essential part of the parcel's title. The same law regarding precedence of claims that applies to deeds also applies to surveys: The earliest recorded claim that is legally valid and accurate will take precedence over all subsequent surveys of abutting parcels that share boundary lines.

An excellent source of legal and technical information on surveys and boundaries is *Evidence and Procedures for Boundary Location* (Brown et al. 1981). Not only does this book touch on the practical aspects of locating and verifying boundaries in the field, it also describes the legal precedents for accepting boundary markers as historically valid and true. If you anticipate entering a boundary dispute, a review of the relevant principles presented in that book will yield valuable information about the strength or weakness of your claim.

## Rectangular Survey System

The rectangular system, or *government survey,* is used in thirty states, mostly west of the Allegheny Mountains. It is not used in states of the original thirteen colonies, in northern New England, nor is it valid in West Virginia, Kentucky, Tennessee, or Texas. As discussed earlier in this chapter, the rectangular system was devised as a means to quickly lay claim to public lands after the Revolutionary War, and to facilitate subdivision of those lands for a fast-growing and westward-moving population.

Beginning with the eastern border of Ohio, the country is divided into north–south running lines—*prime meridians* or *principal meridians.* A prime meridian is intersected by east–west running *baselines*, creating a set of coordinates that become the primary references for all the land around those points. In between prime meridians are *guide meridians*, spaced every 24 miles, and in between baselines are *standard parallels*, also spaced every 24 miles (figure 4.7). Guide meridians are used to adjust for the curvature of the earth, since parallel north-south lines will eventually converge at the earth's poles, and the area in the grid will get smaller and smaller.

Each 24-mile by 24-mile block is subdivided into townships that measure 6 miles on a side. Townships are arranged in east-west rows that extend north and south of a baseline. A north-south column of townships extending east and west of a prime meridian is called a "range" (see figure 4.7). The first township north of the

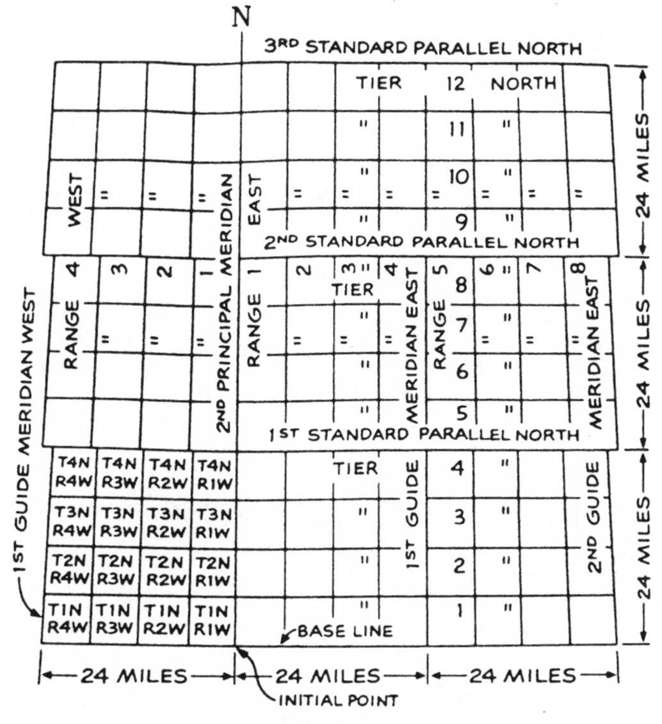


*Figure 4.7. The rectangular survey system is based on the intersection of a principal meridian and a baseline. At 24-mile intervals from the principal meridian, a guide meridian is located to correct for curvature of the earth. The resulting grid is further divided into townships, which measure 6 miles on a side.*

baseline is called "township 1 north," abbreviated T.1N. The first range west of the prime meridian is called "range 1 west," abbreviated R.1W. In this way, every township is referenced by its east-west row number and range relative to the intersection of a prime meridian with a baseline.

Townships are subdivided into thirty-six "sections," which are 1 mile square (figure 4.8). Note that sections are numbered beginning in the northeast corner of the township. Each section is 640 acres, and all subdivisions of a section refer to the half- or quarter-section where the parcel is located. Further subdivisions have similar references, so it is easy to place a rectangular parcel of almost any size in a section of a township (figure 4.9).

Because parcels are rectangular, with lines consistently bounding in cardinal directions, knowing the location also tells you something about the size of the parcel and the length of each of its boundaries. The location of the parcel is also its legal description

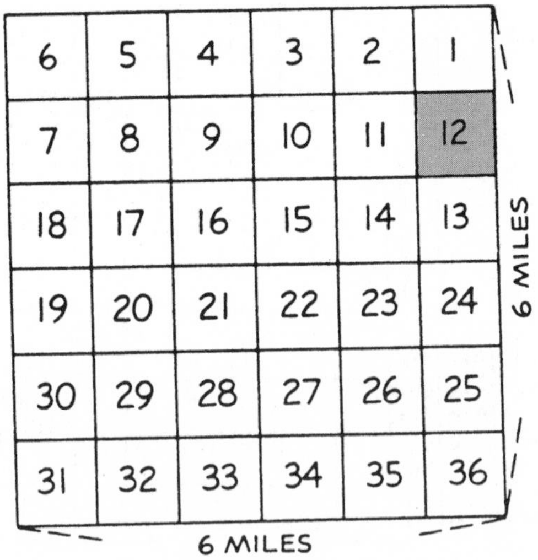

| 6 | 5 | 4 | 3 | 2 | 1 |
|---|---|---|---|---|---|
| 7 | 8 | 9 | 10 | 11 | 12 |
| 18 | 17 | 16 | 15 | 14 | 13 |
| 19 | 20 | 21 | 22 | 23 | 24 |
| 30 | 29 | 28 | 27 | 26 | 25 |
| 31 | 32 | 33 | 34 | 35 | 36 |

6 MILES

6 MILES

*Figure 4.8. Each township is identified by its coordinates relative to the intersection of the principal meridian and the baseline. A tract, township, or tier is defined by east-west lines, and the position east or west of the principal meridian is known as the range. Each township measure 6 miles on a side and is composed of thirty-six sections. Each section is 640 acres.*

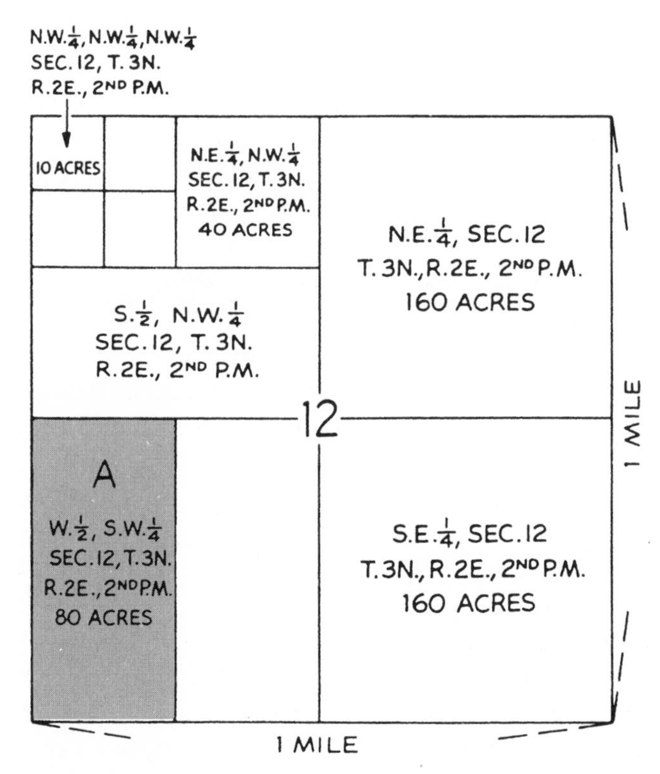


*Figure 4.9. This is Section 12, Township 3 North, Range 2 East of the Principal Meridian. When a section is subdivided, the resulting parcels are rectangular. A parcel's location within the section also provides a legal description for the property, and the size of the parcel. For instance, a half-section is 320 acres.*

in the deed. In many counties where rectangular surveying is used, parcels are mapped relative to one another, with each owner identified, and the information is published in a booklet called a Plat or Tract Book. Many timber buyers and foresters looking for clients use the local "Plat" to solicit business.

## Using a Survey to Locate Boundaries

It is always a good idea to know where your boundaries are, especially if you intend to do work on your land. Some owners make "walking the boundary" an annual event. It is easy to do. All you need is a copy of the survey or a map with bearings and distances, a compass, and your feet to pace distances. An aerial photograph, available through your county USDA Natural Resource Conservation Office (formerly the Soil Conservation Service), is also useful.

A common distance measure of surveyors—both in the rectangular system and in metes and bounds—is the "chain," equal to 66 feet. The chain is a useful unit, since 80 chains, or 5,280 feet, is equal to 1 mile. Also, 10 square chains is equal to 1 acre, or 43,560 square feet. To estimate your pace, lay out a course of 66 feet and walk it four or five times, counting the number of times you plant your right or left foot (i.e., a single pace is two steps, not one). Use a comfortable but deliberate gait, remembering that you will be walking in the woods when counting paces. Average your number of paces in a chain, and you are ready to go. If you have a fairly recent survey, you may need to convert feet into chains before going into the woods.

An important thing to keep in mind is this: A survey is always done as though the land were perfectly flat. To cover the distance of a chain on an incline requires an extra pace or two to account for the change in elevation. The steeper the slope, the larger the correction. A surveyor uses special instruments to account for uneven terrain, but for following your boundaries it is necessary to know only that the steeper the incline, the more paces you will need to account for a single chain on the horizontal. After locating a corner the first time, it is useful to make notes about the surrounding terrain so it will be easy to locate the next time.

New technologies for orienteering, such as hand-held receivers that use the Global Positioning System (GPS), are fun to use for walking boundaries and reasonably accurate for following such a course as between two corners (figure 4.10). The U.S. Department of Defense, however, manages the satellites that provide GPS data, and it has incorporated a random error into the signals, ostensibly to fool the guidance systems of hostile missiles. The error, called "selective availability" by the military, limits GPS accuracy to within 325 feet horizontally and 500 feet vertically, about 95 percent of the time. The error is slightly greater than this 5 percent of the time, so GPS cannot be used to establish accurate fixes on boundary corners, but it can get you back to within sight of

*Figure 4.10. A portable Global Positioning System (GPS) is an inexpensive and easy method of orienteering around your property.*

the house or some other landmark you have programmed into the unit.

If you see flagging tape, painted stakes, or other markers along the boundary during your walk, it is a sure bet your neighbors are planning work on their side of the line or are having the boundary resurveyed. Survey stakes usually mark the position of the surveyor's equipment when taking measurements, not the actual property boundary. Although a wooden stake may serve as temporary monumentation, the surveyor will eventually establish a more permanent marker. Check with your neighbors when you see evidence of a survey along a common boundary, and do not assume stakes found on your side of the line (as you know it) is an attempt to claim a piece of your land.

## Boundary Disputes

Regardless of the system of survey, when people share boundaries, disputes are not uncommon. Because of the problems inherent with metes and bounds, disputes are more common where that system is used, and much more difficult to settle. In more than a few instances, a buyer thought he or she was acquiring "130 acres, plus or minus" only to discover after an abutting neighbor's survey that the land is only 95 acres. Almost more troubling than the instant loss of assets is the owner's realization that he or she has been paying taxes on someone else's land. Usually the only thing an owner in these circumstances can do is to hire a surveyor (and possibly a lawyer) to refute or validate the abutter's claim. Taxing authorities will not refund taxes paid in error, on the grounds that it is the responsibility of the owner to ensure a parcel's title is perfect and valid. Neither will the "new" owner refund the taxes, on the grounds that the "previous" owner enjoyed the benefits of the land. Such a dispute is an easy way to lose many years of carefully planned and executed work in a stand of timber nearly ready for its first harvest.

The best way to resolve a dispute is to get a second opinion. If two surveyors agree, there is nothing to dispute, and you are better off forgetting it. If you have a substantial investment in the timber, the new owner may be willing to work out a deal whereby you can retain timber rights for a specified period of time. The new owner, though, is under no obligation to do so and concessions are agreed to out of a sense of fairness. Threats and demands borne from your outrage at the unfairness of a boundary dispute that goes against you may only sour the new owner, making him or her unwilling to work with you. Bear in mind that even a small portion of something is worth more than all of nothing.

## Posting Land, Trespass, and Personal Liability

When you post your lands, you are giving public notice that your property is off-limits except for those to whom you have authorized access. Many owners assume that placing signs along the border of a property is sufficient to give notice. Most states, however, have strict laws that govern the proper posting of land—laws that may prescribe size and color of the sign, distance between signs, and even the size of letters on the signs. Some states require the name and address and/or phone number of the owner. Also—and this is where many owners fail to effect a proper and legal posting—your state probably requires formal notification with the town or county clerk. Failure to notify the municipality of your posting may subvert your efforts to keep people off your land or limit your ability to prosecute trespassers. Always check with the municipality to learn the rules of posting before setting out markers. In some jurisdictions a proper posting must be renewed each year, and sometimes fees are due.

If you are new to an area, bear in mind that local residents may resent the posting of your lands. Even though it is your right to control access, people in the community may have hunted on your lands, hiked there, or gathered berries for generations. A new owner is under no obligation to continue to provide access to those people, but barring them from the property may cause resentments, which could lead to such problems as vandalism. Before posting, find out who uses your lands and decide if those uses are something you can live with. Get to know people who come onto your land, encourage them to do so, and they will become even more protective of your property than you are. Usually, so long as you do not charge for access, your liability to guests is no greater than your liability to uninvited visitors.

Some owners who post their lands assume it is the best way to avoid personal liability for injuries sustained by uninvited visitors. Although that is a point worthy of clarifying by calling the Attorney General's office in your state, the facts are that posting probably will not provide any special protection from personal liability. This is especially true if you know of hazards on your property, or if you have created hazards to deter access. For example, if you know the location of an old well hole and a trespasser falls into it, you may be liable unless you have made a reasonable attempt to mark the hazard or eliminate it. A deliberate hazard might be an unclearly marked berm or ditch across a road you do not want 4-wheel-drive vehicles to access. An injury sustained by an uninvited driver of a 4-wheeler—even one acting carelessly—may be judged at least partially your fault even though the person was trespassing.

There is also something known as an *attractive nuisance*. This is a feature on your property, such as a tree house, a pond, a cave, or some other feature that is attractive, especially to children or minors. Generally, you must provide extra protection where there are attractive nuisances. Also, some states have laws dealing specifically with this type of trespass. If anything on your property might qualify as an attractive nuisance, check it out with local authorities, and be sure your postings are proper and legal to protect against liability claims from this type of situation.

The concept of liability is based on society's assumption that there is almost never any such thing as an accident. Someone is ultimately responsible, and the degree to which an individual or company is found to be responsible for losses is a measure of liability.

Liability is always an issue in the acquisition of real estate. It is impossible to obtain a mortgage from a bank without first producing evidence of insurance that includes protection against personal liability claims. Many homeowner policies extend coverage to woodlands routinely or in a special clause known as a *rider*. Check with your insurer and seek written clarification on the extent of coverage as it applies to forest lands. Some state woodland owner associations offer special liability insurance tailored to the types of claims a forest owner is apt to encounter.

It is impossible to obtain total immunity from potential liability, short of avoiding all interactions with other humans. However, besides insurance, you can defend against claims in other ways. Always disclose known risks clearly and in writing, and have people who use your woodlands accept those risks in writing and indemnify you from losses they may sustain (indemnification is discussed further in chapter 6). Another way to avoid liability is to be sure contractors and others who work on your land have insurance, including personal injury insurance for themselves and their employees. This type of insurance is known as workmen's compensation insurance, and it is required of logging contractors with employees. Request written proof of this insurance, and if you have questions about the coverage, ask your insurer for an opinion.

# Chapter 5

# Managing and Using Forest Land

Most people acquire forest land for reasons that have little to do with resource management. Periodic timber sales, roads, creating or maintaining wildlife habitats, and other forest management–related activities are not often factors new woodland owners take into account. They are more concerned with the practicalities of siting a new home, locating water and septic, and creating a landscape near the house. After a few years of getting to know the land, the prospects for different resources become more evident. Then, many owners begin to investigate ways to use forest resources and, for some, a plan begins to unfold.

Many things in life do not require a great deal of planning; use of forest resources is not one of them. Woodland owners who launch into timber sales and other activities without advance planning are apt to express dissatisfaction with the results. Thus, the purpose of this chapter is to describe a process for planning the use and management of forest resources.

## Establishing Forest-Management Objectives

Sooner or later, a woodland owner learns that owning a forest property is not an entirely passive proposition. Much needs to be done even if the owner's goals are to leave the forest undisturbed by human hands. Although it is true that an aggressive manager with maximization of resource values in mind has more to do than a passive manager who wants nothing more than to enjoy the natural beauty of forests, both have responsibilities and must put effort into ensuring their respective goals are met. A major difference between active forest managers and passive managers is the degree to which they manipulate forest resources. The motivations of both types of managers may seem worlds apart, but they must have something in common to realize their separate goals: easily articulated objectives.

The most common question forest managers ask of new clients is, "What are your objectives?" This usually precedes a walk through the woodlands, with the manager describing various opportunities. Too often, however, the question remains unanswered. Very few woodland owners can easily say what it is about forest land that makes owning it so appealing. But lacking a well-thought-out answer, the opportunities become the objectives, and the owner relinquishes a degree of control to the manager. In other words, it is left up to the manager to decide what is important, and the owner accepts those judgments. Because managers by training and experience know what is good for the forest, what's wrong with following their advice? Nothing, as long as it reflects your concerns and beliefs about the forest and it meets your long-term goals for the land.

Most woodland owners, though, lack clear objectives, even though forest managers agree objectives are a necessary prerequisite to owning and managing forest land, because objectives control actions (figure 5.1). A woodland owner without them is like a ship without a rudder. Forest-management objectives help identify possibilities and a means of realizing opportunities. They also help to assess success or failure, and to channel resources and effort to areas needing the most attention.

Well-stated objectives also describe the purpose of forest ownership by framing the owner's interests within the context of the forest's resource capabilities. It makes no difference if an owner is utilization oriented or otherwise; clear, easily articulated management objectives are essential to all forest owners. Such a simple question, "What are your objectives?"; yet most owners are at a loss for an adequate answer. This is not surprising, because even a greatly simplified approach to defining objectives results in hundreds of different combinations.

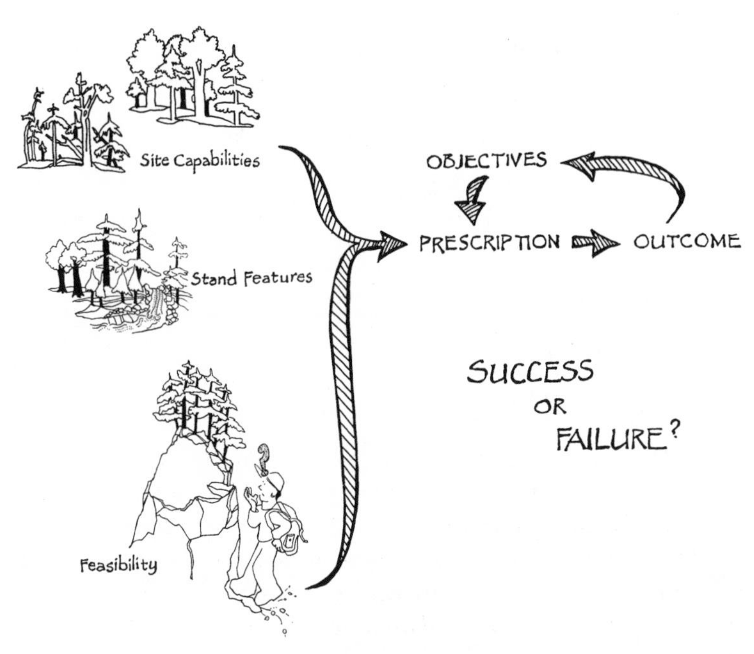


*Figure 5.1. Forest-management opportunities and the owner's objectives control actions in the forest. Success or failure of a particular action is a function of objectives.*

Consider the case of a woodland-owning couple who are asked to state their top three forest-management objectives. To make it easier for the couple, objectives are defined two ways: first, on the basis of forest resource components, such as timber, wildlife, recreation, and aesthetics; second, on the basis of the human benefits the couple want to maximize, namely, income production, long-term investment, and personal satisfaction. Furthermore, assume that each forest resource component must be related to a human benefit, which is often the case anyway. For example, the couple might agree that one of their objectives is to manage timber resources for long-term investment.

The scenario described above results in a matrix with forest resource components down the side and human benefits across the top (figure 5.2). Because most resource/benefit combinations are not mutually exclusive—in fact, the couple could strive to satisfy all possible combinations—the couple needs to evaluate the extent to which certain combinations are more important than others:

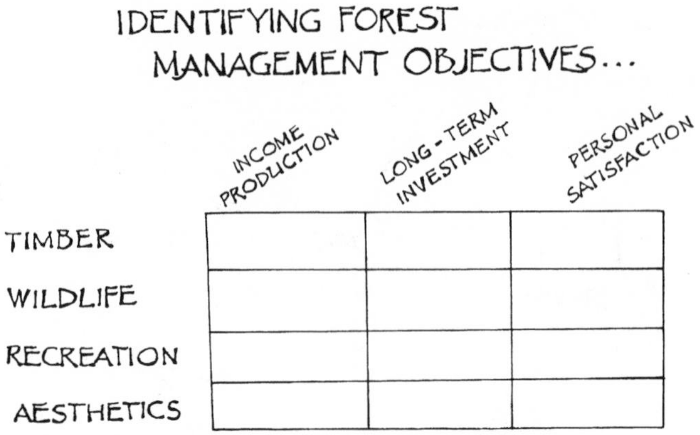

IDENTIFYING FOREST MANAGEMENT OBJECTIVES...

| | INCOME PRODUCTION | LONG-TERM INVESTMENT | PERSONAL SATISFACTION |
|---|---|---|---|
| TIMBER | | | |
| WILDLIFE | | | |
| RECREATION | | | |
| AESTHETICS | | | |

*Figure 5.2. Articulating forest-management objectives is often difficult. If an owner chooses the first three objectives as combinations of human benefits across the top and resource benefits down the left side, so that the cell representing a combination that is of primary importance gets a 1, the combination of secondary importance gets a 2, and the combination of third importance gets a 3—and all other cells are zero—1,319 other choices are excluded.*

They must assign priorities to their objectives. (Remember, this scenario is intended to demonstrate the difficulty of establishing objectives. For most owners, the difficulty arises when one priority obviates the possibility of another priority. The example used here is intended to illustrate this phenomenon, not to suggest that forest-management priorities are always mutually exclusive. In fact, creative managers find the opposite to be true.)

Asked to identify the three most important combinations, such as "timber for long-term investment," "wildlife for personal satisfaction," and "aesthetics for personal satisfaction," the owners number the corresponding cells in the chart 1, 2, or 3. These are their top three objectives to the exclusion of everything else, so all other cells have a zero. How many other sets of objectives have they excluded? Astoundingly, there are 1,319 combinations other than the one they have selected. (The problem is solved by using matrix algebra. The owners have selected one combination out of 1,320 possibilities—12 x 11 x 10 x 1). No wonder forest owners have a difficult time stating objectives.

By putting a 1 in the "timber/long-term investment" cell, a 2 in the "wildlife/personal satisfaction" cell, and a 3 in the "aesthetics/personal satisfaction" cell (and 0 in all the others), the owners in this example have identified themselves as semiaggressive forest managers, willing to invest time and effort to fulfill a fairly active management strategy. Conversely, a matrix with combinations concentrated in the bottom, right-side identify a more passive approach to management (figures 5.3 and 5.4).

The real advantage to viewing forest-management objectives as a matrix of resources and benefits is the ease with which one can state objectives once the difficulty of making selections is resolved.

IDENTIFYING FOREST MANAGEMENT OBJECTIVES...

| | INCOME PRODUCTION | LONG-TERM INVESTMENT | PERSONAL SATISFACTION |
|---|---|---|---|
| TIMBER | | | |
| WILDLIFE | | | 1 |
| RECREATION | | | 2 |
| AESTHETICS | | 3 | |

| | INCOME PRODUCTION | LONG-TERM INVESTMENT | PERSONAL SATISFACTION |
|---|---|---|---|
| TIMBER | 2 | 1 | |
| WILDLIFE | | | 3 |
| RECREATION | | | |
| AESTHETICS | | | |

*Figure 5.3. In the first matrix, the owners choose wildlife for personal satisfaction as their primary objective, recreation for personal satisfaction as secondary, and aesthetics for long-term investment as tertiary. In the second matrix, the owners have established an emphasis on managing timber, which will take priority over other values but not exclude them.*

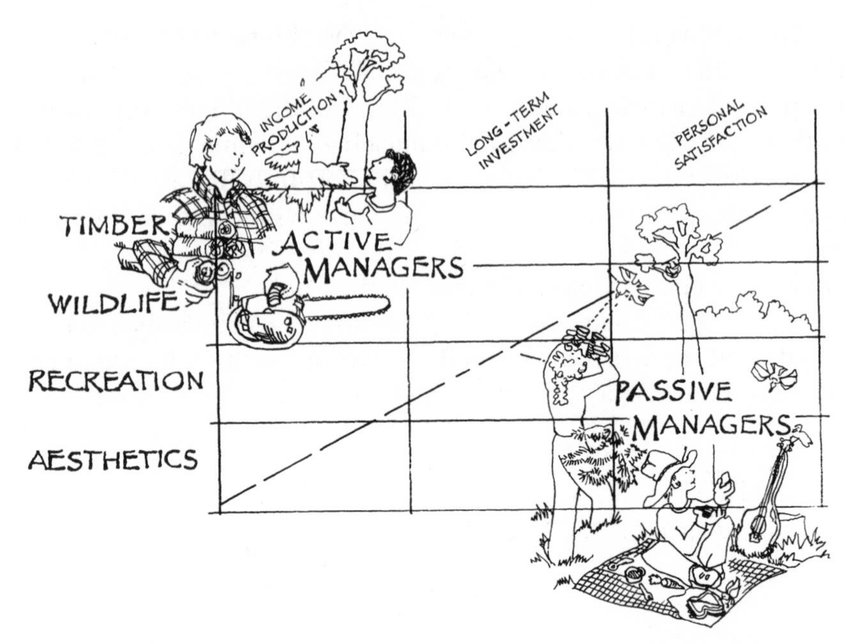


*Figure 5.4. People who choose combinations in the upper left side of the matrix are committing to more active management than people who choose combinations in the lower right. Even passive forest owners need clear objectives.*

For instance, the primary objective for the couple in our example might be stated as follows: Manage timber resources on long rotations to maximize long-term investment. The secondary objective is to create and maintain a diversity of habitats for wildlife, especially songbirds; and the third priority, preserve forest aesthetics. Once the priorities are established, a hierarchy dictates that the secondary and tertiary objectives are subordinate to the primary objective. First and foremost is managing the timber resource for long-term investment. Wildlife habitat and aesthetics are subordinate. How does this affect management decisions? In an area designated for long-term timber investment, the manager will make accommodations for wildlife but only to the extent they do not interfere with the primary objective. The manager always has three things in mind: timber investment, wildlife, and aesthetics. There are many circumstances where all three are easily accommodated, but when conflicts arise, the hierarchy prevails. Fulfilling a primary objective does not preclude wildlife and aesthetics, such as in this case. Rather, the primary objective defines the context or bounds within which other activities are allowed.

Forestry professionals are usually happy to render an opinion on management strategies, and their role is crucial when it comes to recognizing resource opportunities. But the decisions, and ultimate responsibility, for management activities lie with the owner. Clearly stated objectives greatly facilitate decision-making and help to ensure that forest-management activities, like timber harvests, meet expectations and provide satisfaction.

There are instances where the objectives matrix described above does not apply. For instance, if you recently acquired woodland that has been cutover, your options are limited and the best course of action is probably no action. Even if the matrix does not work in your circumstances, it is still a good idea to have objectives.

## Forest-Management Planning

When owners have forest-management objectives in hand, the next step is to develop a plan. The forest-management plan is a document that describes the natural resources on a tract of land. It also includes recommendations detailing how resources can be used to provide a sustained mix of benefits in keeping with the objectives and interests of the owner. Implicit in all forest-management plans, for passive and active managers alike, are the following three elements:

1. A clear statement of objectives.
2. A description of the resources, which may be either brief, verbal descriptions for passive managers, or detailed, quantified resource inventories for more active managers.
3. A chronology of major management activities, prioritized according to the owner's objectives and the capabilities of the resources to meet those objectives. The plan documents resource management opportunities and constraints to achieve the mix of benefits required by the owner. It is not solely a timber management plan.

While an owner with more passive interests in forests may get by with a broadly generalized plan, an active manager needs more information to make good decisions. It is usually only active managers who plan, though, most often in anticipation of timber sales. Regardless of their goals, all forest owners should have a long-term (ten to fifteen years), written plan for the land.

There are as many different types of forest-management plans as there are professionals who prepare them. It is impossible to say exactly what is needed in each instance, and there is no one plan to fit all. To a lesser or greater degree, however, a plan should include the following ten elements:

1. The owner's management objectives
2. Maps of the property and important resource areas
3. Boundary description
4. Forest inventory data
5. Site and stand descriptions
6. Timber management recommendations
7. Local forest product markets
8. Potential for nontimber benefits
9. Other management recommendations
10. Chronology of major management activities

For the owner whose primary objective is "aesthetics for personal satisfaction," all ten elements are not necessary. However, even if the owner has no intention of managing timber resources, the plan should at least document the fact that there are timber resources on the property. Such an owner would need elements 1, 2, 3, 5, 8, 9, and 10. To demonstrate the importance of the plan elements, regardless of an owner's objectives, consider the importance of each to the total plan:

*Owner's Objectives.* Besides consideration for resources and benefits, as discussed earlier in this chapter, you need to think about how long you intend to own the land, and how and to whom it will be disposed when ownership is relinquished. For instance, if you believe land will appreciate during your tenure, and one of your goals is long-term investment (i.e., speculation), think about what a new owner will look for; such elements as good roads, large healthy trees surrounding good potential building sites, and other features that will be attractive to a buyer. In contrast to this scenario, your goal may be to pass on a forest-management legacy to children and grandchildren. The objectives you establish today may one day be fulfilled by your heirs. Regardless of your long-range goals for the land, it will be more valuable with clear objectives and a good management plan than it will be without one.

*Maps.* Many different types of maps (or mapped information) can be included in the plan. The purpose of maps is to show the location of the property and to give readers a picture of where various resources are located. The level of detail varies according to the owner's objectives. Useful map information includes topographic data showing the elevation and relief of the land, soils data, habitats, timber stands, roads and trails, human-made features (such as wells and stone walls), and springs and water courses. A well-designed and clearly drawn map is an excellent way to get the big picture of the forest. Gathering map data, however, and drawing and reproducing maps are expensive.

Some foresters use sophisticated computer-aided mapping techniques known as Geographic Information Systems, or GIS. When map data are read by a computerized scanner, or digitized, the user can generate maps that include as many different layers as necessary (figure 5.5). The real advantage of GIS is the ability to ask "What if . . . ?" questions. For instance, a computer-generated map composed of geologic features, topography, drainage, and soil types would help with road placement and design. The user inputs the ideal conditions for a road, then looks at the various alternatives. Once an acceptable alternative is achieved, the road is laid out in the woods. Another advantage of GIS is that once the base map is scanned and digitized, it is a relatively simple matter to print maps. Data layers can be added as necessary.

Geographic Information Systems are relatively new and many foresters do not use them. Instead, they overlay a simple base map with Mylar, using as many layers as they have data or features to

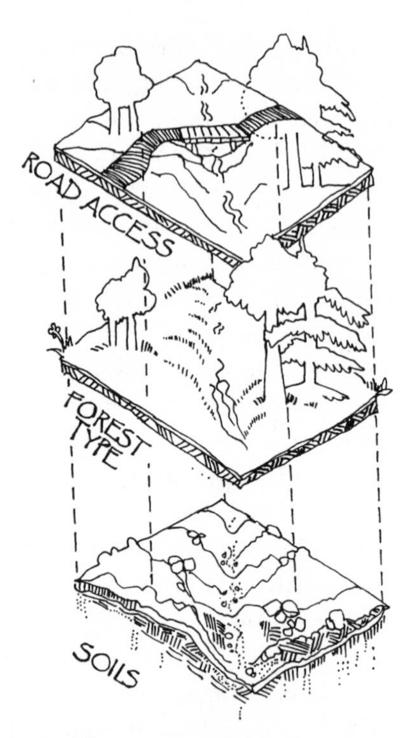


*Figure 5.5. Resource data scanned into a computer using a Geographic Information System (GIS) can layer data in different ways to help the manager make decisions.*

show. Most base maps start with a topographic map, then a data layer on soils, followed by forest vegetation and other layers.

*Boundary Description.* The forest-management plan should serve as a repository for copies of all legal documents pertinent to the forest. This includes a copy of the most recent survey and/or the deed (or at least the boundary description from the deed). Having this information on hand, in one place, makes it easy to locate if it becomes necessary. It is especially important for timber buyers or others doing work on the property to know where the boundaries are (see chapter 6).

*Forest Inventory Data.* The need for detailed forest inventory data is proportionate to the amount of timber management the owner expects to do and the current and potential timber values. An active manager with valuable timber needs fairly detailed inventory information, while an owner with more passive objectives does not. The plan should document timber potential even if the current owner is not interested.

A detailed forest inventory requires the technical expertise of a forester. The forest is divided into areas of similar structure and/or species composition. Each area, or stand, is "cruised," using sampling methods to measure some, but not all, trees. Because a forester measures only a fraction of the trees, statistics are used to verify reliability of data. The timber inventory is only an estimate of volumes, which may be greater or less than actual volume. A good inventory will always show the error of the estimate. In other words, an inventory summary might state: "There are 5,500 board feet per acre, plus or minus 1,500 board feet." Without an estimate of error, such as described, inventory data are of little value.

Generally, the more valuable the timber, the more detailed the inventory and the lower the error of estimate. Stands of low-value timber do not warrant the expense of an excessively precise inventory. It is not unreasonable, though, to expect errors of 10 percent or less in valuable stands, and of up to 30 percent in less valuable stands.

A good inventory also includes information on the age of trees and on growth rates. Knowing growth rate is useful for planning silvicultural treatments and for "growing" the stand at tax time to figure how your cost basis in timber changes after some has been harvested (chapter 8).

Inventory data are commonly summarized on a stand-by-stand basis in the form of stand and stock tables. A stand and stock table gives a detailed picture of what the stand looks like. If one of your top three objectives involves timber management, statistically valid inventory data are invaluable.

*Site and Stand Descriptions.* A "stand" is defined as "a collection of trees that is sufficiently uniform in species composition, structure, condition, or age to be distinguishable from surrounding areas" (Smith et al. 1997). A wildlife habitat biologist may refer to a stand as a "cover type." The collective management of all stands, or cover types, on the property in question is forest management. The stand description may include stand and stock tables, information about the site such as slope and aspect, soils, management limitations, and a measure of productivity. The predominant species of trees and plants are identified, and the structure of the stand is described. The forester also notes any disease or insect problems, or other risks from such factors as fire or wind. Knowing the condition of the stand allows the manager to estimate the risk of losses from natural causes during the life of the plan. The emphasis in this part of the plan is on potential.

*Timber Management Recommendations.* If timber is an important resource for the owner, the plan will include recommendations for all stands that have been identified. Each stand has an objective in keeping with the overall management objectives of the owner. For instance, an owner may have wildlife as a primary goal for the forest, but a particular stand may be well suited for long-term timber management. The stand objective describes the desired structure, age of trees, or other conditions that control when the harvesting is to commence. It is based on two things: (1) the owner's overall goals and objectives, and (2) the stand's capability to provide benefits in proportion to the owner's objectives.

A timber management recommendation usually includes a *prescription,* which is a statement describing actions to achieve the objectives for the stand. The prescription should also describe the expected outcome of a treatment. The ability to predict outcomes is the basis on which a prescription is founded. For instance, if a prescription is intended to secure new regeneration of a particular species, there needs to be a guideline to use after the treatment to assess success or failure. In the event of failure, the prescription should identify a Plan B, so the original objective is fulfilled. One of the principal faults of timber management prescriptions is not allowing for failure, with procedures to correct the situation.

The timber management recommendations also describe constraints that may affect treatments, such as erosive soils, steep access roads, or other similar conditions.

Finally, since implementing timber management prescriptions is often the most disruptive of forest-management practices, the forester should describe the risks and trade-offs of a particular treatment. The potential benefits should always exceed the risks. In timber management, it is better to do nothing than the wrong thing.

*Local Forest Product Markets.* A good forest-management plan will include information about local product markets for timber and other resources. Some consultants are reluctant to include this in the plan because they feel the information is proprietary. Nevertheless, if you are paying someone to develop a forest management plan, you are paying for information, and information about local markets is usually critical to implementing the plan. All aspects of the plan should stand alone and not necessarily require the services of the person who prepared the plan once it has been delivered. Although it may cause a breach of trust to take a plan developed by a consulting forester and use the information either by yourself or with the services of others, the facts are that you own the plan and can do as you please. An itemization of local markets is not a standard element of most plans, so you may need to request it.

*Potential for Nontimber Benefits.* The plan should describe recreational opportunities, sites of archaeological significance, scenic vistas, special habitats, unusual or valuable geological deposits, areas for hunting and fishing, and any other nontimber resources on the property. Many forest owners anticipate passing land to children, so it is also a good idea to identify areas where they may want to build. Anyone who has built a house in the forest knows a good site is a very valuable resource. Your children and grandchildren will celebrate your good foresight.

The plan should also identify critical landscapes, such as ridge lines, that require special treatment to preserve the aesthetic appeal of the forest. It should also disclose special considerations for protecting threatened and endangered plant and animal species that might use, or reside on, your land.

*Other Management Recommendations.* Unique management opportunities, such as small-scale hydroelectric generation, a rare species of wildflower, or an opportunity to improve or maintain habitat for wildlife should be described in the plan. Also included are recommendations on roads and trails, boundary maintenance, stream crossings, or any other activities that need to be undertaken to effect the plan.

This section should also reference any local ordinances, state regulations, or other laws that might affect forest-management activities. For instance, all states have rules to protect water quality during harvesting. Generally, it is the responsibility of the forest owner to ensure those rules are followed. The exact nature of local best-management practices to protect water quality should be spelled out in the plan or clearly referenced so a copy of the rules is easy to locate.

*Chronology of Major Management Activities.* One of the most useful aspects of the forest-management plan is a chronology of activities that need to take place. The owner knows where to start and can concentrate on tasks that yield the greatest benefits. Some foresters are reluctant to give too much detail for fear the owner will no longer request services. Nevertheless, the plan is like a road map: you should be able to read it and use it without someone else's help. When you pay for forest-management planning services, be sure the document is clear on what needs to be done. Also, planning is not a one-time event. A plan usually has a life of ten to fifteen years. After that, it is time to reassess your objectives, analyze what worked and what did not, and consider the future. If timber management is a high priority, it is also time to reinventory.

## Working with Forestry Professionals

Generally, anyone who makes a career of providing services to forest owners can be characterized as a forestry professional. There are major differences, though, in credentials, motives, and abilities. Most states have no laws that govern credentials, and anyone, regardless of education or experience, can choose any title he or she feels appropriate. The purpose of this section is to provide a clear distinction among the many different types of people who offer services to woodland owners. In a few states, the titles people use and their corresponding credentials are controlled by law. Check with your local state extension forester to see if any laws apply.

The Society of American Foresters (SAF) is a professional organization that represents approximately eighteen thousand individuals connected with forest land management. Regular membership is extended to people who have completed a four-year bachelor of science or graduate degree in forestry from an SAF-accredited college or university. The society has minimum educational requirements in a wide range of subject areas that an institution must fulfill to maintain its accreditation. A "forester" in the eyes of SAF is someone who is eligible to be a member, not necessarily a member in good standing. On the other hand, there are many individuals who meet SAF membership criteria who are not members.

It is generally accepted in the profession that a forester is someone who has completed a baccalaureate, master's, or higher degree in forestry. An individual claiming to be a forester should be able to produce a diploma that has the word *forest* or *forestry* in it. In other words, a bachelor of science degree in biology would not meet SAF guidelines (unless the person is also a full member of SAF). Where credentials and titles are not controlled by law, the issue of who uses the title "forester" (or makes references to services that only a forester can provide) is the subject of much debate when forestry professionals come together. If you are looking for forestry services in connection with planning or implementing elements of an existing plan, be sure the person you are working with has the necessary credentials: a university degree in forestry from a recognized school and at least some experience in the areas in which you need help.

### *Public versus Private Foresters*

It is important to understand the distinction between foresters in the public sector and those who are private. Public foresters include most research and teaching foresters, and state and federal land managers. They also include *extension foresters*—mentioned frequently in this book—whose job is to provide education programs for owners, managers, and users of forests, and, finally, *service foresters*, usually employed by the state natural resource agency. The job of service foresters, though it varies from state to state, is to provide one-on-one technical support to woodland owners. Of the public foresters, a woodland owner is most apt to use the services of extension foresters and service foresters. Generally, the mission of these public services is not to provide free services but to help woodland owners make good decisions about their woodlands. For instance, a local service forester will visit with a woodland owner, offer an assessment of opportunities, and provide guidance on local services. The service forester is also a valuable source for second opinions. However—and this is a key point—the public service forester cannot in any way represent your interests. As a public servant, he or she cannot act as your agent or offer any advice or guidance that favors one source of private services over another. A public forester's job would be in jeopardy if the forester were promoting the services of one provider to the exclusion of others.

There are two types of private foresters: those who work for the forest product industry and those who sell services to (or act as agents of) private woodland owners and others.

The first type is known as an *industrial* or *procurement* forester. Also known by the term used in the forest industry, Landowner Assistance Program (LAP) forester, the important thing to remember when working with these foresters is that their first responsibility is to their employer—the mill that is interested in your timber. The individual may be personable, competent, and concerned about your objectives and your woodlands, but there is no ethical way he or she can act as your agent or hold your best interests above those of the employer. When you understand this (and the forester has explained the relationship to your satisfaction), an industrial forester and the implied relationship with a local forest products industry can be an appropriate and rewarding relationship. But don't forget: They are protecting their company's interests, not yours. Also, their services are usually provided in anticipation of a timber sale, for which the company will at least have the right of first refusal, although this is not always a requirement. The industrial forester should not request or accept any direct payments from the forest owner for services. The exceptions to this are instances where the company allows the employee to moonlight as a consulting forester, in which case the forester is presumably in a position to act as your agent and hold *your* interests above all others. Where there are industry ties, know who you are working with and request full disclosure of all potentially conflicting relationships.

The second type of private forester is known as a *consulting forester*. The consulting forester, either as an individual or as an employee of a company that provides consulting services, is in business to represent only the interests of the woodland owner. Either as a private contractor or as an agent (see chapter 6), a consultant's first responsibility is to his or her client. Occasionally, a consulting forester will have a logger or a mill as a client. If such a relationship poses even the remote possibility of a conflict with your interests, the consultant should disclose the relationship and allow you to decide how to proceed.

Consulting foresters offer a wide range of services, but they are usually experts in forest inventory, management planning, and timber sale administration. If you require special expertise in such areas as wildlife habitat, forest finance, and taxation or estate planning, you may need to seek a consultant competent in these areas. Also, all consultants are not equal. Do not assume that a university diploma automatically makes someone a good source of forestry expertise. Always ask for and check references before hiring any professional services.

Another difference between an industrial forester and a consultant is that the woodland owner pays the consultant for services. Payment can take many different forms, but a percentage of gross income from timber sales is common. This method of payment assumes a timber sale will take place, and for this reason it is often an inappropriate way to pay for services. Most consultants are open to other methods of payments, methods that do not rely on immediate timber sales. (The business nature of an owner's relationship with consulting foresters is discussed in more detail in chapter 6.)

Aside from technical competence, some of the most important qualities you should look for in a consulting forester are an ability to listen and to communicate. Sometimes professionals are reluctant to explain their actions to clients on the grounds that the client does not have a technical background sufficient to make explanations short and simple. If there is something you want to know, you deserve an explanation. Many consultants will offer an initial visit for little or no charge. This is a great opportunity to get to know each other and discover any communication barriers. Choosing the right forester can be difficult. There are many excellent, highly competent consultants who are poor communicators. Most areas of the country have enough local consultants to give you a choice. Even though working with a consultant is usually a long-term relationship, you should always consider working with someone else if the relationship deteriorates. And remember, the consultant may be the expert, but you are the boss.

## *Logging Contractors*

A logging contractor is the person or company responsible for extracting timber. The contractor may or may not have forestry credentials, and he may be the buyer, the buyer's agent, an independent contractor, or—in some circumstances—the woodland owner's agent or partner. The logger is usually either an independent contractor or the buyer.

Although many states in recent years have initiated education programs for loggers, there are no third-party accredited training programs in the United States for logging contractors. For most loggers, their credentials are the sum of their experiences. The majority of logging contractors are competent, honest, and easy to work with. The public's negative image of loggers is the result of the incompetence, greed, and dishonesty of a few individuals. One need only ask around to discover the renegades in the logging community. If you learn something about a logger that would compel you to shun his services, it is best to keep your reasons to yourself rather than to confront the logger with what you know. You are under no obligation to explain your decisions to anyone. Unless you know exactly what you are doing and trust the logging contractor implicitly, you should first seek the advice of a consulting forester or the local service forester, although the service forester's ability to render an opinion in this matter is limited.

Recent changes in forest-management practices are causing changes in logging communities. Today, there are fewer fly-by-night operations and fewer loggers. Where loggers used to have work lined up for a few months at a time, now they experience a backlog of a year or more. A good logger is in high demand.

## Tips on Working with Professionals

The following are a few points to keep in mind when working with forestry professionals:

- Separate the sales pitch from the expertise. Know when information is being used to inform and when it is being used to sway.
- Do not trivialize the forestry professional's expertise. Caring for forests is not rocket science, but there is more to it than meets the eye.
- Be cautious with contractors sporting new, highly sophisticated equipment. Chances are the debt on the equipment is enough to make you catch your breath, and good practices can readily fall victim to the bill collector.
- "If an offer appears too good to be true, it probably is."
- Be clear about who is paying the expenses of a professional working on your land. Forestry consulting fees usually include ordinary and reasonable expenses (phone, travel, supplies, etc.). An industrial forester's expenses are paid by his company, and a logger is usually responsible for his own expenses. If you have questions about unusual or costly expenses, clarify the matter up front.
- Reserve the right to terminate your relationship at any time for any reason, understanding that you must pay for services already provided.
- If you have no intention of working with a local forest products company that is willing to provide you with forestry services, you should locate a private consulting forester instead.
- Never proceed with something you do not fully understand. Ask the professional for an explanation or seek answers from the local service forester or extension forester.
- Ask for credentials and do not hesitate to seek documentation. Also, remember that credentials of any kind do not guarantee performance.

- If a personal relationship begins to evolve, try to keep it separate from the business relationship. Being close friends with a local logger's family is one thing, but making business decisions on the basis of your friendship is another.
- Discover any agency relationships that may not be clear. For instance, a logger who appears to be an independent contractor is actually an employee of a local mill.
- Do not agree to work with someone just because you know the family, or because the person is related—an uncle, cousin, or brother. This type of arrangement fails more often than it succeeds, and you are probably better off locating someone else. Any ill will from such a decision is quickly forgotten, while memories of a bad experience with a family member can last a lifetime.
- If you have a bad experience with an individual, it is best to sever the relationship and keep the details to yourself. If you must complain, or if someone asks, be careful about what you say and to whom, and how you say it. The reasons for this caution have to do with threats of lawsuits, forcing you to substantiate your claims.
- When you have an extremely positive, rewarding relationship with a forestry professional, take the time to write a letter of appreciation. Everyone enjoys congratulations for a job well done, and the letter will go into the portfolio of testimonials from satisfied clients.

## Chapter 6

# Forest-Management Contracts

Forest management is all about controlling forest assets. Whether it is for sustained cash flow from periodic timber sales, recreation, wilderness, long-term investment, improved wildlife habitat, or a combination of these and other objectives, management implies control. For most forest owners it is impossible to effect forest management without entering into contracts. Ironically, it is usually the prospect of entering into agreements with others that frightens uninitiated woodland owners away from forest-management opportunities. They reason it is easier to do nothing and avoid conflict. But even the most passively involved forest owner will need to enter into contracts from time to time.

The three essentials of good management are these:

1. A clear idea of what is important—your objectives and an ability to share that vision with others.
2. Competent local foresters, wildlife biologists, loggers, and others whom you trust.
3. Good written contracts to ensure your concerns and interests are understood and protected in dealings with others.

In all states, with the exception of Louisiana, laws governing contractual relationships are enumerated in the *Uniform Commercial Code* (UCC). Developed in the early 1950s, and adopted in most states by the mid-60s, the code is composed of ten articles whose purpose is to unify rules of conduct in business relationships and thereby facilitate interstate commerce. Covered here is only a small portion of the UCC that relates specifically to the types of transactions forest owners are apt to encounter during their tenure with the land. Most of the technicalities of contracts as outlined in the UCC fall outside the scope of this book and the code varies slightly among states. Some excellent, introductory-level textbooks on business law are available that present most details of the UCC, primarily through case studies (Frascona et al. 1984).

For most people, the subject of legal contracts conjures a negative image of human interaction: We have been led to believe contracts are a necessary evil to keep others honest and to avoid being cheated. We see contracts as a means of forcing behavior, to compel people we do business with to act according to our conditions with the threat of being sued if they do otherwise. Contracts are a buffer to the outside world; they are immutable, sacrosanct, and a good way to pin somebody down, or be pinned. Contracts, many of us believe, are the only way to make deals with people we do not trust. They are the moral equivalent of using the carrot as a stick. And, perhaps the greatest misunderstanding of all—only lawyers can develop contracts. None of this is true, and the person who sees legal contracts as a means of effecting any of the above conditions is a victim waiting for a catastrophe.

Contracts are simply a means of formalizing communication between people, and whether we know it or not, most of us enter into contracts more than a few times each week. From implied contracts at the self-service gas station to oral agreements, such as with the kid who mows the lawn, almost all circumstances that involve an exchange of promises—either expressed or implied—will form a contract.

Virtually all contracts related to forest management should be in writing. A good rule of thumb is this: Anytime you make an agreement with someone to do something that will take more than a day to complete, get the details in writing.

Finally, view contractual relationships in a positive context. Some people in the forestry community take offense when offered a contract to sign, saying, "A handshake has always been good enough for me." If you trust the person and intend to do business, follow up the handshake with a letter or memo outlining the points on which you have agreed. If problems arise down the road, you will be glad to have something in writing, and if certain conditions are met, the letter itself may be viewed as a written contract.

## Introduction to Contracts

By definition, a *contract* is a promise, or set of promises, exchanged between two or more parties that creates a legal obligation to perform. It is important to understand the distinction between an agreement and a contract. An *agreement* may be a manifestation of mutual assent between parties, but it does not necessarily constitute a legal obligation to perform. The difference between an agreement and a contract is that the exchange of promises represents value to the parties, value they would not otherwise be required to share if not for the terms of the contract.

All contracts, written or verbal, have the following four basic elements:

1. Evidence of agreement, or a manifestation of mutual assent, is essential to the formation of a contract. All parties must clearly understand the terms effecting a "meeting of the minds." A good contract also clearly demonstrates the definite intentions of the parties to perform the contract and that promises and commitments have been exchanged. The condition of mutual assent implies that willingness and understanding are basic to a contractual relationship.

2. Consideration is value exchanged for the commitments of others. Usually thought of as money, consideration is anything of value given to secure the promises of others, including another promise.

3. Legal capacity of the parties to a contract implies the court will acknowledge a participant's ability to create a contract with others. For instance, a person judged mentally incompetent does not legally have a capacity to contract, presumably because there is no way the first condition—a meeting of the minds—can be met. In most states, minors (under the age of twenty-one in all but a few states) have only a limited capacity to contract. If you

enter into a contract with an individual who has only a limited capacity to do so, you may be forced to uphold your end of the bargain while the person of limited capacity is let off the hook. This is important to keep in mind for owners who commonly hire young people to help around the tree farm.

4. Legality. The contract must encompass only actions, terms, and conditions that are not illegal. Any form of consideration that is apt to be viewed by a court as an agreement to perform an illegal activity may result in a void or voidable contract. For instance, a promise to steal a piece of equipment and deliver it to another party in exchange for promises from others (whether those promises are legal activities or not) is an illegal contract.

If any one or more of these four elements is missing, a contract may be judged void, voidable, or illegal. The first and third elements are more conceptual and obscure than the elements having to do with consideration and legality. It is usually easy to ascertain consideration in a contract but not always as easy to determine if there was a meeting of the minds. A contract delivered by a timber buyer to an elderly and infirm woodland owner, who may be incapable of understanding all the terms, could be voided simply on the grounds that there was no meeting of the minds.

Contracts can be implied or express, and express contracts can be written or oral (figure 6.1). An *express contract* means the terms agreed to were certain and stated by the parties, while an *implied contract*, as the term suggests, means that by their actions (or inaction) a contract is assumed to exist between the parties. For example, an owner watching someone paint his or her equipment shed cannot refuse to pay by arguing that no contract exists. The owner's inaction—failure to stop the other party from working—may be viewed as an implied promise to pay. Although each type of contract varies in the degree to which terms are specified (or assumed), an implied contract has all the force of law that an express contract does.

The Uniform Commercial Code requires that some contracts be express and written to prove their existence. For instance, contracts for the sale of land or for the sale of an interest in land (such as timber and other resources) need to be in writing. Contracts

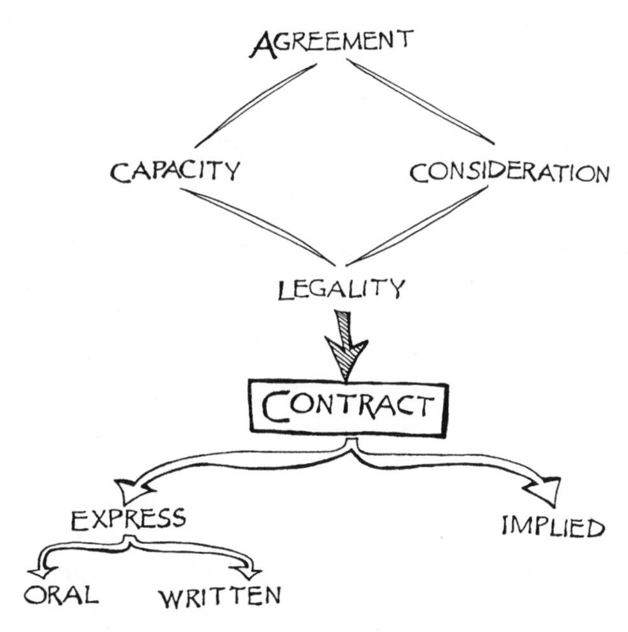


*Figure 6.1. The four essential elements of a contract can be stated in specific (express) terms—oral or written—or implied based on the actions or inactions of the parties. Generally, forest-management contracts should be expressed and in writing.*

between parties for the performance of services generally do not need to be in writing. (The UCC as it applies to the sale of timber and other assets from land is discussed later in this section.)

In its simplest form, a contract can represent a single promise between two parties. For example, you promise to pay a forester $500 for an undefined inventory of your woodlands. This is called a "unilateral contract" because it involves one promise—to pay $500—dependent on completion of an activity. When the inventory is delivered, the forester can expect payment regardless of the fact that there were no performance expectations to judge the acceptability of the inventory data. It is called a unilateral contract, because the forester has not made any promises. This is a perfect example of a bad contract, because there are many different types of inventory, from a cursory walk-through to a detailed, statistically validated survey of timber volumes. Also, if time of performance is not stated, reasonable time is implied, which may be considerably longer than you think is reasonable. Contracts for services should always take the form of a bilateral, written contract—with an expiration date—unless you know the service provider extremely well.

There is no such thing as a standard contract in forestry practice. If the four conditions are met, a legally binding contract can be developed on the back of a napkin. The term *standard* usually means a form commonly used by the party offering the contract for signature. In forestry transactions, it is often the timber buyer or the provider of services who will offer a contract to the woodland owner. Among merchants, however, it is commonly the seller who first offers the terms of a contract. The buyer then either accepts the terms, rejects the offer, or submits a counteroffer. A forest owner offered a contract can extract elements to which he or she agrees and draft a new contract for signature by the other party. This is regarded as a counteroffer, and depending on the complexity of the terms, consideration and reconsideration of counteroffers are a form of negotiation.

Nothing should prevent you from developing your own contracts. Should a discrepancy arise, a decision as to who is at fault will be based solely on the terms of the contract and all the events that have occurred since the contract was formed. For these reasons it is important that the text of the contract cover all aspects of the relationship between the parties. It is not necessary to have an attorney draw up a contract for it to be binding on the other party. But it is usually well worth the cost to seek an attorney's opinion on a contract, particularly if it is someone else's. The cost of this service can usually be recovered at tax time.

The following are a few other points to keep in mind before signing forest-management contracts:

- Include an "indemnification clause" to limit your liability for the actions of others. When the other party accepts an indemnification clause, he or she agrees to "save and hold [you] harmless" for damages or injuries that might arise as a result of the contract. There are limits, of course, to which this type of clause will apply. In other words, it probably will not provide full protection if you are actually liable for an injury or loss.

- Specify an expiration date but include a clause that allows extensions for good cause, at your discretion.
- Eliminate any wording or reference that appears to be outside the scope of the contract.
- Get the other parties to agree to settle disputes using arbitration (see chapter 10). A common practice is for each party to select an individual to help arbitrate and then the parties (or the arbitrators) agree on a third arbitrator. The decisions of the arbitration panel are final. This can be an excellent alternative to going to court.
- Request a performance bond if any of your assets are at stake.
- Include a paragraph that clearly states the outcome or products you expect from the contract. Do not hesitate to do this; the other parties will tell you if your demands are unreasonable.
- Always have someone witness the contract—an individual who can be easily located even after a few years, preferably a notary.
- If you decide to sign a contract developed by someone else, have your lawyer look at it, but explain your concerns, so the lawyer knows what to look for.
- Finally, include a paragraph stating that you will write a letter to the other parties upon full performance indicating the contract is complete and you are satisfied with the outcome. The letter formally states that in spite of any other documents, statements, implied circumstances, or facts to the contrary, your contractual relationship is complete. In other words, the contract has been successfully executed.

## *Purchase of Services*

A *service provider* is someone you contract with to perform a task. The key element to understanding your relationship with the provider is this: You are purchasing the result of services rendered and not the right to control how the services are provided, except as specified in the contract. Also known as a *work order*, the contract for services can be express or implied, oral or written. It can be unilateral (I'll pay you $1,000 to survey my land), or bilateral ( . . . if you promise to complete the survey within thirty days). Most written work orders are bilateral, with the time-frame and other performance conditions clearly spelled out. If it is your intent to control how services are provided, you either need a detailed work order or you may want to consider developing an agency authorization, which is discussed later in this chapter.

There are many different types of forest-management–related service providers you can expect to encounter—from realtors to heavy equipment operators, surveyors, accountants, and others. The most common service provider you are likely to encounter is a consulting forester. Private consulting foresters can provide a wide range of services, from locating boundaries to mapping and inventory. They can also administer timber sales and help with some of the business aspects of management. Best of all, by having no conflicting ties with the wood-using industries, a private consulting forester can represent your interests and yours alone. In other words, under certain circumstance described below, a forester can act as a fiduciary.

When entering into service contracts, a clear understanding of the product is essential. For example, a contract to develop a forest-management plan could result in a few pages of general statements about the land coupled with a hastily sketched map, or it could result in a comprehensive and detailed document with inventory data, management options and consequences, and a series of carefully drawn maps. In a contract that does not define product expectations, both documents are apt to be viewed equivalently. If there were a discrepancy that ended up in court, testimony on what is fair and reasonable given the facts would weigh heavily in the court's judgment.

Another point to bear in mind about work orders or service contracts is to know what you are *not* purchasing. An agreement to develop a forest-management plan is not necessarily an agreement to begin implementing the plan. One of the most common errors of this nature involves boundary surveys. The owner's concept of a survey is a carefully drawn map with distances and directions, and a survey area with marked boundaries. Unless requested in the contract, most surveyors will not mark the boundaries as part of the survey.

Contracts for services are a useful method of purchasing services as needed. Because they must be carefully negotiated each time a new contract is developed, however, and may not provide you with an adequate degree of control, they often fall short when working with a professional over a period of years. If you have located a forest manager you trust, someone who is easy to speak with and who understands your concerns, a far more useful and satisfying contractual relationship may be one that authorizes the manager to act in your behalf. Clear communication, flawless rapport, and a high degree of trust are essential prerequisites.

When working with independent contractors focus on the following:

- Once the contract is signed, you do not have a right to control the actions of the service provider except as outlined in the contract. Be clear about any conditions you want met.
- Payment for services can be arranged many different ways, but a portion of the payment (up to one-third of the total) should be held until the contract is fully executed.
- Expenses to complete services are commonly paid by the provider, except as outlined in the agreement. For instance, a client may be charged for phone calls and travel expenses, but not for secretarial time or copying expenses. Include a clause in the contract that spells out agreed-on expenses.
- Be sure the contractor has workmen's compensation insurance for his or her workers and that any subcontrators are also covered. This is a form of liability insurance that protects the employer and the clients from claims of workers injured on the job. Also, check to see if the contractor carries general liability insurance and that it is adequate to cover damages or injuries to others that might occur during execution of your contract. Some owners will request that the contractor's insurer list the owner as co-insured during the life of the contract. Check with your insurer.
- Clearly state that the contractor is not an employee, and that

nothing in the contract should be construed as an employer–employee relationship.

- Pass liability for compliance with applicable laws and regulations to the contractor, who should be willing to attest to the fact that all laws and regulations will be followed. Bear in mind, though, that most laws of this nature are written so that the landowner is ultimately liable for compliance.
- If the service provider will be handling money for your account, be sure the money is held in an escrow account and that you will be provided a complete accounting when the contract is finished.
- Generally, a service provider is not necessarily your agent, so be sure the wording reflects the exact nature of your relationship. Some work orders or agreements will refer to the service provider as your agent. Unless that is your intent, change the wording.

## *Agency Authorizations*

Common law, which varies from state to state, describes an *agent* as a person with a fiduciary responsibility to someone he represents. A fiduciary has a legal duty and authority to act for the sole benefit of another. The relationship can be implied or express, oral or written; as long as the agent is doing things solely for the benefit of the parties he or she represents, it is an agency relationship. For the sake of discussion, the entity whom the agent represents is known as the *principal*. The parties with whom the agent enters into contracts for the benefit of the principal are known as *third parties*. In forest-management situations, the woodland owner is the principal, the agent may be a consulting forester (or anyone the principal authorizes), and the third party is anyone with whom the agent enters into contracts for the principal (figure 6.2).

How does an independent contractor differ from an agent? You buy only the results of a contract with an independent contractor. If a person is acting as your agent, you have a right to control his or her actions. An agent can also enter into contracts that are binding on the principal while avoiding personal liability for the contract

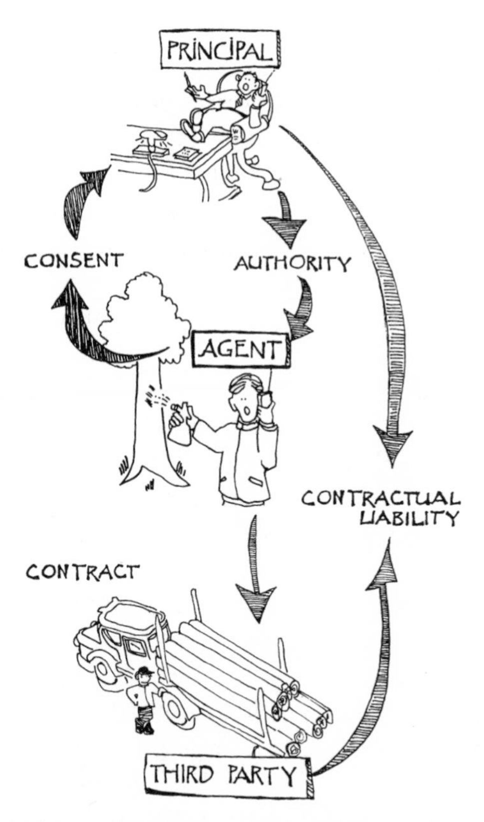


*Figure 6.2. The relationship between the principal and third party created by the authorized actions of the principal's agent. The agent must be acting with the authority of the principal, and the third party must know this to be true.*

(assuming it is perfectly clear to the third party that the agent is acting in the principal's behalf and the principal's identity has been fully disclosed).

A written agency authorization is also known as a *power of attorney*. When the scope of an agent's activities are strictly defined, it is called a *limited power of attorney*. If the limited power of attorney is also a contract between the agent and the principal, the law of contracts applies. Setting up an agency relationship is an extremely powerful way to work with professionals, but it is not without risk. The scope of authority vested in the agent by the principal must be clearly spelled out, and an agency authorization should usually have an expiration date of twelve months or less with a provision to extend if the principal so desires.

Because you have the right as principal to control the activities of your agent, each of you has expectations regarding the other's behavior. It is generally accepted that the agent's duties to the principal include performance, loyalty, obedience, care, accounting, and information. The principal's duties to the agent include compensation, compliance with the terms of the agency, reimbursement for expenses sustained by the agent and indemnification—to "save and hold harmless" the agent in his or her dealings with third parties in behalf of the principal.

A contract for an agency relationship, besides outlining the scope of the agent's authority, should identify a form of consideration (i.e., money) to make the contract binding. In this case, consideration takes the form of a *retainer*, a fee paid to the agent that can be credited against compensation due the agent when acting in your behalf. The retainer is like an advance and most likely will not include the total compensation the agent can expect.

The contract for agency has two parts: (1) The limited power of attorney granted to the agent by the principal, or the exact scope of the agent's authority, and (2) the contract, which besides satisfying the four elements of a legal contract, lists the procedures by which the agent is expected to fulfill his or her duty to the principal.

It is not unreasonable for the contract to require regular written communication from the agent, or for the agent to seek prior written approval before entering into certain types of contracts—such as timber sales. The contract should also clearly describe the special nature of what is otherwise viewed as an employer–employee relationship. Unless it is your intent to hire the person as an employee, you need to state clearly that the agent is not an employee but a person you have entrusted to act on your behalf according to the limited power of attorney. This is tricky and may require some legal advice from local attorneys who have experience in this area.

Here are some other points to consider:

- List the powers of the agent as exclusionary; that is, if not specifically listed, the agent does not have authority to act without your written consent.

- If the agent is empowered to make purchases for you, notify merchants to that effect.
- If the agent is empowered to sell timber for you, notify local log buyers.
- Reserve the right to terminate the agency contract at any time for any reason, but be willing to pay what is due the agent for services rendered prior to termination but not yet accounted for.
- If you are concerned about legal incapacitation during the agency, you may want to have your lawyer create what is known as a *durable power of attorney* by including the words "This power of attorney shall not be affected by the subsequent disability or incapacity of the principal," or wording to the same effect.
- An agent cannot be used to act in behalf of someone who is already legally incapacitated. For instance, if the principal has only a limited capacity to contract, the agent must operate under the same limitations.
- Anyone can act as an agent for another, even a minor. But remember, the law of contracts applies: You may not be able to enforce performance by a minor.
- If the agent exceeds the scope of his authority, it is the agent, not the principal, who is responsible for any contracts or claims, unless the principal ratifies the agent's actions.

## Forest-Management Leases

A lease is a type of contract that grants exclusive property rights to a person (the *lessee*), usually for a specified period of time. The property can be personal property (goods) or real property (land). The difference between the two is that personal property is tangible, something the lessee can possess and use, while real property deals with more intangible rights the lessor is willing to grant the lessee, such as the right of access. In northern New England, owners of mature northern hardwood forests will often lease the right to gather sap in the spring for maple syrup production. A lease is a way for an owner to allow others to use property while enjoying the benefits of income—the rent a lessee is willing to pay for those uses. While the owner (lessor) still holds title to the property and is responsible for property taxes, leasing is a great way to help defray the costs of owning forest land. The laws governing contracts apply to leases.

Forest management leases usually require the lessee to install improvements, so it is not uncommon for the lease to extend five to ten years or more. For example, a sugar-maker lessee bears the cost of tailoring sap-gathering tubing to an area that is to be tapped. It would be impractical to do this if the lease were good for only one year. Generally, if the lessee's activities do not require an expenditure of capital, lease life can be shorter. The longer the lease, the more the lessee should be willing to pay.

In some parts of the country, mineral exploration companies will approach forest owners to lease the right to explore for oil and gas and other valuable minerals. Those leases are usually long-term, and the rent is paid up front as an inducement to sign. There is also a "royalty" provision if the company locates exploitable minerals on the property. At face value, a mineral exploration lease looks like a good deal. The problems arise when the lease is exercised and access roads are poorly designed and other site disturbances are not corrected before the crew moves on to the next site. Be wary of mineral exploration leases. Always seek a legal opinion before signing one, and don't sign if you are not completely comfortable with the deal. If there are valuable minerals in your valley, one or more of your neighbors probably will sign, and eventually you will hear about it.

Observe the following points when signing leases:

- Be sure you first own the rights you are leasing. For instance, some forest owners do not own mineral rights on their lands. You cannot sign a mineral exploration lease if you do not own the right to do so.
- The lease will create an imperfection on the title. An astute lessee will record a long-term lease with the town or county to protect his or her interests if the property goes up for sale.
- Always reserve the right to cancel the lease for cause with ade-

quate and reasonable notice to the lessee. Spell out the circumstances that would cause you to cancel the lease.

- The lessee should be willing to post a bond or provide a deposit to cover the cost of damages or to compensate the owner for failures on the part of the lessee.
- A lease to use your woodlands should be nonassignable in the absence of prior written approval of the owner. In other words, the lessee cannot pass rights to another person without your permission.

## Timber Sale Contracts

According to the American Forest and Paper Association, two to three million private, nonindustrial forest owners will enter into contracts each year for the sale of forest products. Less than a third of those transactions involve any type of professional forester, and a majority of sales do not involve written contracts between buyer and seller. Yet the law requires a written contract if the sale involves more than $500 worth of timber to be cut by the seller, or—the more likely scenario—the timber is to be cut by the buyer. In the first instance, timber is treated as "goods" under the Uniform Commercial Code (UCC 2-107); in the second instance (the buyer does the cutting), it may be treated as a sale of an interest in land. Both conditions require a written contract to prove the existence of an agreement between the buyer and seller. Aside from the fact that it is plain good business to "get it in writing," when you are selling timber a written contract is required in every state except Louisiana.

One of the only practical ways to change or improve forest conditions for wildlife, recreation, or timber production is through periodic timber sales. The sale provides income to help cover the costs of owning land, and by manipulating the vegetation an owner can influence forest conditions for many different uses. The trick is to generate income while improving forest conditions. In most parts of the country, harvesting timber solely for the sake of income production is not a valid strategy. Most successful managers see timber as a by-product of forest management that aims to improve investment, create wildlife habitat, and sustain other less tangible but no less important forest values, such as biodiversity.

Always look for four parts in a timber sale contract (figure 6.3). Usually, the first paragraph identifies the buyer, the seller, and any agents. This may seem obvious, but it isn't always. For instance, you may sign a contract with a timber buyer, and then a week later a logger you have never met shows up and immediately violates the spirit of the contract. The logger may be the buyer's agent, but he is more apt to be an independent contractor. He is responsible only to the mill who bought your timber, and the mill has purchased the result of his activities, not the right to control his behavior. How do you resolve that situation? The buyer should divulge the names of people retained to harvest the timber and should also be willing to show you a copy of the agreement with the logger (the work order). The work order should convey exactly the same conditions you have placed on the buyer, with it being the buyer's responsibility to supervise the activities of the contractor. Anyone who will be involved in the sale should be identified.

The what, where, and when of the contract identifies the species to be harvested, the products (i.e., sawtimber versus pulpwood or other products, such as poles for pilings), the precise location of the sale (preferably with a map showing the boundaries of the sale area and surrounding property boundaries), and the duration of the contract and a listing of circumstances that may dictate when harvesting should be suspended, such as during wet ground

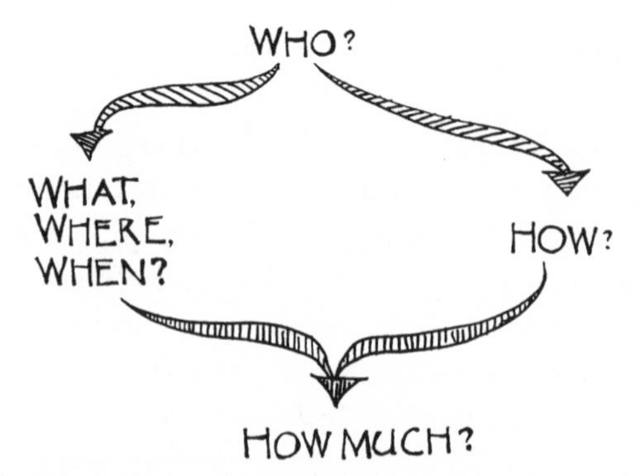


*Figure 6.3. Look for these critical elements in a timber sale contract.*

conditions. A timber sale contract should always have an expiration date, with provisions to extend it for cause.

Often a timber sale contract will identify only the principal products, such as sawtimber (the main trunk of the tree), but not the by-products, such as cordwood in the tops of trees. If it is the intent of the buyer (or the timber harvester) to remove wood from the tops for sale, it should be listed as a product in the contract and you should be credited for its value. From an income tax perspective, you are better off to sell the tops in standing timber than to sell the wood from tops after the trees have been felled.

It is your responsibility to ensure the buyer knows where the timber is located. And it is usually the seller who proposes access routes to harvest areas. To ensure compliance with the terms of the contract, it is a good idea to confirm that the buyer is well oriented to the sale area. If, after the sale, there are claims of trespass from an abutting neighbor, your efforts to ensure that the buyer was well informed will help protect you from liability.

The how of a timber sale contract details all of the conditions you require of the buyer. It addresses such conditions as treatment of slash after harvest, reseeding roads and landings, access routes, and specifications for skid trails and haul roads. It might also specify the type of equipment the buyer will use on the site, and any other special conditions you require, such as protection of wildlife den trees and preservation of stone walls and cellar holes. If the landing is near your house, you should be aware of the fact that loggers often start work before first light, and they work on weekends, too. If you like to sleep in on the weekends, be sure to specify limitations on work hours.

Always retain the right to suspend the operation for any cause, but especially during ground conditions that could be injurious to tree roots. In the Northeast, every spring has two to four weeks when frost is leaving the ground. During this period, all but extremely sandy soils are susceptible to damage from heavy equipment. Known as mud season, it is a time when logs are scarce and loggers go without work. You should reserve the right to suspend harvesting during those conditions, or you will run the risk of damaged timber, washed-out roads, polluted streams, and angry neighbors.

Place as many special conditions on the sale as you need to feel comfortable. Remember, however, that each condition requires special attention from the buyer, which means more time and less money. So be willing to accept less for the timber from a buyer who is willing to respect and follow the conditions you require.

The contract should identify timber volumes and the means by which those volumes are to be measured. There are at least six major methods used to estimate the volume of boards in logs, known as log rules, and they all differ. Each region has its own traditional log rule. In the Northeast, we use the International 1/4-inch rule. In the Midwest, the Doyle rule is common, and on the West Coast, they use the Scribner Decimal C rule. Local variations exist for each of these rules, so it important to know how the timber is being measured. Sometimes the final harvest volume is not known when the contract is negotiated. Even so, the contract should provide some hint as to how much timber is to be cut and removed from the land during the course of the contract.

The method of payment is an important issue for the seller and buyer to resolve. Traditions vary from region to region, but the strategy for the seller is to ensure that payments are prompt and in the correct amount. Prepayments are not uncommon, and if the amounts are more than a few thousand dollars, the seller should require a cashier's check or a bank-certified check. The method of payment often depends on the type of sale, of which there are three: lump-sum sale, mill tally sale, and sale on the basis of shares.

The lump-sum sale is the most risk-free method for the seller. The buyer agrees to pay the seller a lump sum for the right to harvest designated timber, regardless of the actual volume the buyer recovers. It is the buyer's responsibility to ensure the sale volume is accurate, and he or she bids accordingly. The owner, who cannot possibly be held responsible for poor utilization standards of a logging crew or for internal defects revealed at the mill, is paid based on the estimate of volume in standing trees. The contract can specify a payment schedule for large sales or a single payment when the contract is signed and before harvesting begins.

In a mill-tally sale, the buyer agrees to pay the seller so much per unit of wood, the units to be determined when the wood is measured or tallied at the mill. The risks for the seller are the following:

1. Who will ensure the logging crew is maximizing log values in the woods?
2. How will the owner know that logs are being delivered to only one mill?
3. Who actually owns the logs once they leave the property?
4. Who is liable if a logging truck carrying logs to market gets into an accident?

Following are the answers to those questions:

1. There is no way to control utilization standards of logging crews short of constant supervision or some sort of incentive that gives loggers an interest in maximizing log values.
2. One can keep track of log loads, but it is tedious and consulting foresters rarely do it.
3. As far as ownership of the logs is concerned, the title is shared between the buyer and the seller until the logs are first measured at the mill.
4. You may be liable if a log truck carrying your logs gets into an accident. The trick is to pass title of the logs (or at least pass liability) when they leave the landing on your property. Your attorney can help craft a specific clause in the contract to pass title—and responsibility—for logs to the buyer the moment the buyer's truck leaves the property.

A sale on the basis of shares is really a consignment sale. The buyer (usually also the logger) agrees to pay the seller a percentage of the mill-delivered price for the logs. The percentage in the Northeast varies between 30 and 40 percent share to the seller. Of the timber sale methods, this is the most risky for the seller. Why? Because the seller is going into business with the logger, and in doing so is placing complete trust in that person's technical capabilities as a logger and in his or her own business practices as a partner. The same risks as for the mill-tally method apply as well.

Regardless of the method of sale, most woodland owners should seek the advice and guidance of a private consulting forester before selling timber. It is the consultant's job to represent the interests of the owner and to see that all the conditions of the contract are met and that payments are made on time. When a consultant is involved, his fees for administering the sale are usually paid as a commission from the gross proceeds of the sale. This is accepted practice throughout the country. However, the owner should be aware of a potential conflict of interest when a consultant is paid on commission: The more timber sold, the higher the forester's commission, whether it is good forestry or not. Most consultants will disclose this situation before the sale; some consultants do not. It is perfectly acceptable to pay a consultant on the basis of commissions, as long as you understand the implication of doing so. Also, because of the potential for conflict of interest, it is probably impossible for a consultant to hold a limited power of attorney (or even act as a fiduciary) in connection with timber sales. The best situation is to tie payments to services rendered, not to commissions on sales.

Before signing a contract to sell timber, follow these ten steps:

1. Clearly establish your ownership goals and objectives, and the specific management objectives to be met by the sale.

2. Have the harvest area cruised to establish current timber volumes, both growing stock and potential harvest volumes. Also, estimate the growth rate of trees.

3. Develop a prescription—exactly how will forest vegetation be changed and what is the expected outcome?

4. Develop marking guidelines to help implement the prescription.

5. Mark the stand and tally harvest volumes.

6. Identify contract conditions (restrictions) and draw up a sample contract.

7. Solicit bids and show the sale area.

8. Review bids and select a buyer. (Do not reject other bidders yet.)
9. Check the references of the successful bidder.
10. Review the contract conditions with the buyer, amend if necessary, and sign the contract in front of witnesses.

If there is a discrepancy during the sale, the written conditions of the contract are the only source a court will use to identify a breach of the contract. Therefore, it is essential that all conditions are clearly spelled out in writing and that the contract anticipates as many foreseeable circumstances as possible. Once the contract is signed, oral promises mean nothing in court. If you forgot to spell something out and you cannot live without it, put it in writing, have all affected parties sign it, and append it to the contract.

The following are some other ideas to keep in mind when developing a timber sale contract:

- Review the elements of a good contract, described earlier in this chapter.
- An indemnification clause to "save and hold harmless the seller" is essential. Require workmen's compensation and personal liability insurance of the buyer.
- Require the buyer to post a bond or provide a sum of money to be held in escrow to ensure performance of the contract.
- Include a paragraph that requires parties to arbitrate disputes before going to court.
- Payments either go directly to the seller or to the seller's representative (the consulting forester, assuming the forester is acting as a fiduciary). If the payments go to a consultant, be sure you understand how and when payments are to be dispersed to you. Also, be aware of the fact that any payment of more than $300 must be reported to the IRS on the appropriate Form 1099 (discussed in greater detail in chapter 8).
- Do not allow the contract to be assigned to other parties without your permission.

- Require full disclosure of all other parties who may be involved in the contract. For instance, include the name of the heavy equipment operator hired to put in roads.
- Clearly spell out the terms under which the contract will be considered fully executed, and the timing and method for refunding deposits.
- Specify practices that will result in improvements to the land after harvest. For example, have ski trails designed so they can easily be converted to hiking and cross-country ski trails.
- Always reserve the right to terminate the contract for good cause with adequate notice to the buyer.
- When in doubt, seek a second opinion.

Finally, if you are not absolutely comfortable with the arrangements, do not sign the contract. The advantage of dealing with timber is that it will not spoil. You can delay your decisions indefinitely, or at least until you are sure the contract does what you want it to do. If you do not understand any elements of the contract, have your consulting forester and/or your lawyer explain them to you.

# Chapter 7
# Ethics in Forestry Practice

Someone once said ethics are all the things we learn by the age of six, usually on the lap of a parent. It is a matter, we are told, of recognizing right from wrong. We are encouraged to choose a path of right behavior that is standard, not special, and we learn responsibility. In time, we come to recognize the obligation we have to other people and to the world we live in. Our ethics form the basis of a social contract that exists among humans. If not for this contract, human behavior would be unpredictable, chaotic, and violent.

The formation of ethics is a cultural phenomenon. What is acceptable in one culture may be taboo in another. It is almost impossible to predict correct behavior in a foreign culture without hints about what is acceptable and what is unacceptable. The purpose of this discussion is to view forestry and forest management as a culture unto itself and to describe acceptable and unacceptable practices forest owners are apt to encounter.

Most of the ideas presented here were inspired by, or extracted from, *Ethics in Forestry* (Irland 1994), a collection of essays on various interpretations of situations where questions of ethics arise.

## Excessively One-Sided Transactions

A free-market system operates on the premise that both the buyer and the seller are equally knowledgeable about the product and the price. For most woodland owners who periodically sell timber, this is not the case. They know woefully little about the products they sell and even less about fair prices for those products. A forest owner will sell board feet by the thousands without the slightest idea of what a single board foot is. And it is common for a seller to accept the estimates of a buyer without any possible means of checking the accuracy or veracity of those estimates.

When a woodland owner relies exclusively on information from a buyer, it may be an excessively one-sided transaction. A one-sided transaction is not necessarily unethical, but it has the potential to be unless the buyer explains to the seller the one-sidedness of the deal. It is unethical for a buyer to take advantage of the ignorance of a seller even if the seller is acting out of greed. But it is *illegal* to mislead a seller to compel him or her to enter a transaction. Though these statements may seem obvious and irrefutable, in real life they are not. In a capitalist society, we are taught that business is business, and there is no requirement to suffer a fool's folly. Businesses make money with information, and a woodland owner selling timber is entering the wood products business, if only briefly. One can easily argue that it is not the responsibility of the buyer to educate the seller, especially if it means less profit for the buyer. From a strictly business perspective, an uninformed woodland owner is fair game. But if the buyer's advantage is excessive, it is not a fair transaction.

What is the buyer's obligation in this circumstance?—to tell the woodland owner that, as the buyer, he cannot protect the owner's interests and that the owner should seek guidance from someone who can. The act of informing someone of a one-sided transaction is sufficient to remove the specter of unfairness. The buyer is under no other obligation to educate the seller, and the seller is under no obligation to act on the advice of the buyer.

The best way for a woodland owner to avoid an excessively one-sided timber transaction is to employ the services of a consulting forester. The cost of those services is usually more than offset by the financial advantage of the forester's knowledge.

## Conflict of Interest

When a person has an interest in a transaction substantial enough that it does or might reasonably affect his or her independent judgment in acts he or she performs for another, it is a conflict of interest (Barry 1994). For example, in an excessively one-sided transaction, the buyer has a conflict of interest in his relationship with the seller. The conflict arises from the buyer's first responsibility to his or her own interests or those of his or her employer. Another example of conflict of interest is when a person shows favoritism for direct or indirect personal gain. An employee choosing a material supplier because of personal promises the supplier makes to the employee is an example of favoritism and is a conflict of interest. Any use of one's position for personal gain is potentially a conflict of interest.

Offering or accepting bribes or kickbacks as the basis for a business transaction is often a conflict of interest. In forestry, finder's fees are common. It is a payment, usually from a timber buyer to a third party, for information that leads to a timber purchase. Although a finder's fee does not always create a conflict of interest, there are circumstances when the payment or acceptance of those fees is unethical. For instance, if a finder's fee is paid by someone who then represents the interests of the person the information applies to, the payer of the fee is obligated to tell that person about the fee. Otherwise, it could be a conflict of interest.

Another example of a conflict of interest involves related payments from more than one source. An example of a related payment from more than one source is when an agent of the woodland owner accepts a premium (a reward or a kickback) from a mill for delivery of highly valuable logs while also accepting payments from the owner for services. In this case, the agent has an obligation to tell the owner about the other payments.

The best way to avoid conflicts of interest is to fully disclose all the facts of a situation to all the parties involved in a transaction. A consulting forester who provides services in exchange for a commission paid from the gross proceeds of a timber sale has an obligation to explain the potential for conflict of interest to his client. The conflict arises from the fact that the forester has an interest in the transaction (the commission), which might reasonably affect his independent judgment. That is, the more timber he marks, the more he gets paid. The consultant's obligation to his client is to explain the potential for a conflict of interest, even if the consultant never intended to take advantage of the situation. There is nothing inherently wrong with proceeding in a situation where there is potentially a conflict of interest, as long as there is full disclosure to the parties who might suffer unfavorable consequences.

## Code of Ethics

Ethical behavior is largely voluntary. Grossly unethical behavior may be judged illegal. It is not illegal to be unethical but only to the extent that it does not cross into the realm of illegal activities. The continuum between ethical and illegal behavior is unbalanced (figure 7.1), with a broad range of interpretation between ethical and questionable practices. What one person may view as ethical, another sees as questionable. Depending on the facts and circumstances, it may be difficult to judge. As one moves along the continuum from questionable to unethical, it is easier for objective observers to agree on when someone has crossed the line. The distinction between a questionable practice and an unethical one is more obvious; there are fewer shades of gray. The distance between unethical and illegal behavior is very narrow, but the definition is even more obvious. In other words, it is easier for people to agree, because the boundary between unethical and illegal is usually well defined and easy to interpret. A person who behaves in a grossly unethical fashion is probably also acting illegally.

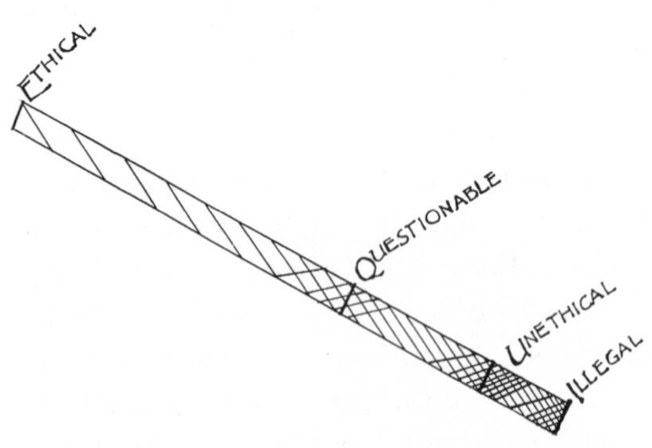


*Figure 7.1. The continuum between ethical, questionable, unethical, and illegal behavior is unbalanced. In forestry, there are many shades of gray between ethical and questionable behavior, but there is almost no distinction between unethical and illegal behavior. Grossly unethical behavior is illegal behavior.*

Recognizing the need to interpret situations in forestry practice that may be questionable or even unethical, such professional societies as the Society of American Foresters and the Association of Consulting Foresters have developed codes of ethics. The codes consist of a list of canons or statements to which all members agree (figure 7.2). The statements are necessarily broad and subject to interpretation, and only rarely is a member formally charged with a violation. An official finding of a violation almost never happens. Most members of these national societies and other similar local professional organizations are aware of their society's code of ethics. But only a few members know the code well enough to paraphrase even one or two of the canons. This is not to suggest that members who cannot recite the canons are inherently unethical. Ironically, usually the opposite is true. Most members of professional societies do not need a code of ethics to determine right from wrong behavior. The codes are more for putting the public's mind at rest than for governing the behavior of the society's members. They also define the boundary of acceptable behavior for those few professionals who operate on the fringe of ethical and questionable practices.

Mollie Beattie, a good friend and now deceased colleague, once said (jokingly), "'Your society [referring to SAF], upon request, will send you a copy of the Code of Ethics, suitable for framing.' The question is, suitable for framing whom?"

The codes are often misused by dishonest individuals who hide behind them or by righteous people pointing fingers. The codes developed by professional societies are necessary, however, and occasionally useful. But steer clear of people who talk about their society's code of ethics rather than behaving in an ethical fashion.

When you work with forestry professionals, as a client you can expect—at a minimum—three things. First, that the nature of your business dealings will be held strictly confidential. It is unethical for someone you do business with to discuss the nature of your relationship, or anything related to a transaction, with third parties. Second, anything other than total honesty and complete loyalty as it relates to your business dealings is unacceptable. It is unethical for a professional to switch loyalties as may be convenient for him or her, and dishonesty is always unethical, whether you are the victim or someone is being dishonest with another per-

*for Members of the Society of American Foresters*

## Preamble

Stewardship of the land is the cornerstone of the forestry profession. The purpose of these canons is to govern the professional conduct of members of the Society of American Foresters in their relations with the land, the public, their employers, including clients, and each other as provided in Article VIII of the Society's Constitution. Compliance with these canons demonstrates our respect for the land and our commitment to the wise management of ecosystems, and ensures just and honorable professional and human relationships, mutual confidence and respect, and competent service to society.

These canons have been adopted by the membership of the Society and can only be amended by the membership. Procedures for processing charges of violation of these canons are contained in Bylaws established by the Council. The canons and procedures apply to all membership categories in all forestry-related disciplines, except Honorary Members.

All members upon joining the Society agree to abide by this Code as a condition of membership.

## Canons

1. A member will advocate and practice land management consistent with ecologically sound principles.

2. A member's knowledge and skills will be utilized for the benefit of society. A member will strive for accurate, current, and increasing knowledge of forestry, will communicate such knowledge when not confidential, and will challenge and correct untrue statements about forestry.

3. A member will advertise only in a dignified and truthful manner, stating the services the member is qualified and prepared to perform. Such advertisements may include references to fees charged.

4. A member will base public comment on forestry matters on accurate knowledge and will not distort or withhold pertinent information to substantiate a point of view. Prior to making public statements on forest policies and practices, a member will indicate on whose behalf the statements are made.

5. A member will perform services consistent with the highest standards of quality and with loyalty to the employer.

6. A member will perform only those services for which the member is qualified by education or experience.

7. A member who is asked to participate in forestry operations which deviate from accepted professional standards must advise the employer in advance of the consequences of such deviation.

8. A member will not voluntarily disclose information concerning the affairs of the member's employer without the employer's express permission.

9. A member must avoid conflicts of interest or even the appearance of such conflicts. If, despite such precaution, a conflict of interest is discovered, it must be promptly and fully disclosed to the member's employer and the member must be prepared to act immediately to resolve the conflict.

10. A member will not accept compensation or expenses from more than one employer for the same service, unless the parties involved are informed and consent.

11. A member will engage, or advise the member's employer to engage, other experts and specialists in forestry or related fields whenever the employer's interest would be best served by such action, and a member will work cooperatively with other professionals.

12. A member will not by false statement or dishonest action injure the reputation or professional associations of another member.

13. A member will give credit for the methods, ideas, or assistance obtained from others.

14. A member in competition for supplying forestry services will encourage the prospective employer to base selection on comparison of qualifications and negotiation of fee or salary.

15. Information submitted by a member about a candidate for a prospective position, award, or elected office will be accurate, factual, and objective.

16. A member having evidence of violation of these canons by another member will present the information and charges to the Council in accordance with the Bylaws.

*Adopted by the Society of American Foresters by Member Referendum, June 23, 1976, replacing the code adopted November 12, 1948, as amended December 4, 1971. The 1976 code was amended November 4, 1986, and November 2, 1992.*

*Figure 7.2. The Society of American Foresters' Code of Ethics.*

son but in your favor. Tolerating unethical behavior is in itself unethical. Finally, you have a right to expect technically sound and acceptable management practices of the professionals with whom you work. It is unethical for professionals to purport technical proficiency in areas where they are unskilled. And it is wrong for a professional to implement practices without knowing if the practices are regulated locally or if other special conditions apply.

## Licensing, Registration, and Certification of Professionals

There is a recent trend for states to control forestry activities and the people who provide services to woodland owners. The purpose of these efforts is to limit the severity of cutting, to ensure forest ecosystems are protected, and to protect the public from people who claim professional competency but have none.

Cutting restrictions and forest protection mechanisms take the form mainly of *notification statutes*, which require the owner to file a plan when harvesting more than a certain threshold amount of timber. To see if your state or local area requires a permit to harvest timber, contact your state Extension Forester (Appendix A). In some locales, permitting is based on access to public roads, on protecting water quality, or even on some obscure (and possibly illegal) local ordinance aimed at controlling timber harvesting. If there are such ordinances in your area, you must comply with them. Generally, the requirements are fairly easy to satisfy. As the landowner, it is your responsibility to be sure your forest-management operations comply with local laws.

Some states are requiring foresters, loggers, and other woods workers to document and maintain credentials to provide the types of services they offer. The statutes take on many different forms, but they are largely intended to help the public understand the extent of a person's competence, and/or to require full disclosure of the professional's relationships with the wood-using industry or others whose interest may be in conflict with those of the forest owner.

A certification program is the least restrictive method of controlling the credentials of people who offer services. Certification usually is voluntary and often sponsored by a professional organization. It commonly requires members to have attained a certain level of professional competence and to make a commitment to maintain that competency. When certification is required by statute, it takes the form of a registration. For a forester or logger to provide services, they must comply with the certification requirements. To maintain the certification, they must usually complete continuing education programs and adhere to a standard of conduct or a code of ethics. If an individual fails to complete the requisite education or violates the standard of conduct, certification is suspended.

Licensing is the most restrictive form of controlling credentials. It is only rarely used, because forestry is somewhat obscure and poses minimal threats to human health, safety, or welfare. Most states reject licensing because of the expense associated with maintaining a board of licensure and on the grounds that only a small portion of the population benefits. The difference between licensing and registration is in the degree to which the controlling authority takes responsibility for licensees. A licensing authority assures the public that people with licenses to practice have achieved a minimum level of competency and that the licensee will perform in an acceptable and professional manner, in accordance with all laws and regulations. Registration does not usually provide the public any guarantees other than of credentials. Regardless of the method—certification, registration, or licensing—each has a standard of conduct by which professionals are expected (or required) to abide.

The American Forest and Paper Association, under its Sustainable Forestry Initiative, has recently asked its member companies to require education programs for its loggers. To comply, a logger must complete a state-by-state defined curriculum within a certain time. Significantly, this is the first time the forest industry has required any conditions on the procurement policies and practices of its members. It is not a certification program, however, and currently no standards of conduct are required of loggers who complete these programs. The SFI education requirement is not a certification program, but it looks like one.

The element that licensing, registration, and certification programs have in common is a standard of conduct or code of ethics that people are expected to abide by to maintain their status. Consumers of professional services must remember that the behavioral standards of certification or licensing are not necessarily a performance guarantee. It is very difficult, almost impossible, to resolve a complaint with a licensing board. Therefore, it is up to the woodland owner to judge a professional's character and to know when and how to enter and break off a professional relationship.

## Environmental Ethics

Many people have come to realize that, besides the social contract we have among ourselves, we also have a responsibility to the earth and to all the other organisms with which we share the planet. Having a sense of environmental ethics implies that consideration is always given to the effects of our activities on plants and animals, and on future generations. Forests are complex, though, and it is not easy to predict the results of our actions. Often, what is good for one species may spell doom for another, or seemingly dire and irreversible outcomes may turn out to be temporary. How then does one know when a practice is right or wrong? The answer is, it depends—on the motivations of the forest owner, on the long- and near-term forest-management objectives, and on the degree to which practices preserve the integrity of the ecosystem. For example, a forest owner who harvests timber solely for production of income without regard to the health and condition of the forest after the harvest is acting unethically; a forest owner who harvests timber for income, but only to the extent that long-term productivity is protected and adequate consideration is given to wildlife species that use the forest, is acting ethically. The irony in these examples is that there may be little difference in the financial outcome for both owners, but one of them is doing the right thing by exercising a little extra care in the interest of other species and of future generations.

Making the right decision is not always easy. Some forest owners believe it is highly unethical to disturb the forest in any way. Yet we need the products of forests and have demonstrated we can harvest timber without sacrificing other values. Human use does not need to be destructive or to obviate use by other organisms. Careful planning with an eye to the future and empathy for other organisms will almost always lead to the right decision. We are fortunate to live in a time when our knowledge of forest ecosystems is expanding rapidly. As our understanding increases, so does our ability to make the right management decisions.

## Tolerating Unethical Behavior

When we tolerate unethical behavior in others, either by looking the other way or by taking advantage of someone else's inappropriate behavior, it is unethical. It is wrong for us to espouse a standard that we live by while accepting lower standards of others; this is especially true if we stand to gain from another's transgressions. For example, it is unethical for a mill that purports to be environmentally sensitive to accept timber from a logger who is notorious for unethical practices. The act of buying logs from people like that is supporting and reinforcing bad behavior, and the mill is just as guilty as the perpetrator.

People associated with forestry are often judged by the bad behavior of a few who act in disregard for any appropriate or reasonable standards. It is not necessarily our responsibility to confront renegades, but it is important to speak out when confronted by people who associate the bad behavior of a few with the entire community. Unless you have facts to support the allegations you could make, however, it is best to criticize the practices rather than the people. Making oral statements of facts that are injurious to the reputation of others is called "slander." There are many ways to criticize bad behavior without taking things to a personal level. Making injurious *written* statements of facts is called "libel." Both are punishable as "torts," which are a noncontractual, civil wrong that results in injury to another person. The truth of the facts in slander or libel is a defense, and the injured party must prove there was intent to cause injury.

Although not nearly so egregious as openly tolerating unethical behavior in others, failure to accept responsibility for one's actions is a distressingly common form of unethical behavior. It is easy to point a finger, to plead ignorance, or to otherwise avoid responsibility. Often a subordinate—the last person hired—takes the brunt of criticism for someone else's failing. Recognize failures to accept responsibility and know that they are symptomatic of behavior that could easily lead to much more serious transgressions. "Acting ethically is acting courageously" (Irland 1994). Taking responsibility for one's actions—especially when it is easy to blame someone else—is acting courageously.

## A Case Study in Forestry Ethics

In Vermont we have a sixty-four-hour curriculum for logging contractors known as LEAP, an acronym for Logger Education to Advance Professionalism. The program is voluntary and covers a diverse curriculum that includes a day-long work-shop on professionalism and ethics. Our approach to teaching ethics includes the use of case studies, where loggers in small groups are asked to review the facts of a scenario and to answer questions about the ethics of various decisions made by the characters.

The exercise is initially intimidating, because the characters and situations are true to life. Eventually, a lively discussion ensues, and the exercises become a favorite part of the workshop for participants. They learn that, although it is not always easy to state why a particular action is right or wrong, instincts can help. They also learn that it is easy to change the thinking of their peers just by speaking up.

Below is one of three case studies we use during the workshop: Read the scenario, give careful consideration to the facts, and think of ways that potentially unethical actions could have been avoided. Then try to answer the questions that follow. An interpretation of the scenario and some answers to the questions follow the case study.

A logger gets a call from a friend who recently heard that his wife's uncle is thinking about selling his timber. The 140-acre woodlot is fairly well known locally, and not much harvesting has been done over the years. Aside from a little fuelwood, there have been no commercial timber harvests on the property for more than sixty years. There may be as much as a million feet of mostly hardwood sawtimber of much better than average quality. The friend suggests to the logger that he give "Uncle John" a call, gives him the phone number, and even offers some ideas on how to sway Uncle John to sell. But if Uncle John does decide to sell, the friend wants "a cut" for making the referral without Uncle John knowing his involvement in the transaction.

The next day the logger parks his pick-up and hikes into Uncle John's woodlot to have a look-see. The area is a beautiful stand of one-hundred-year-old-plus sugar maple and yellow birch. Even a light cut will yield easily accessible, mostly high-quality sawlogs.

The logger calls Uncle John and discovers that he is interested in "selling some timber, but not all of it." With virtually no understory over most of the woodlot, Uncle John expresses concern about "getting a new forest started underneath." The logger suggests a "selective harvest" and Uncle John agrees.

When the logger and Uncle John meet a few days later, the logger agrees to pay so many dollars per thousand for maple and birch, and he also agrees to document log volumes with mill-tally slips. After much discussion, they agree that only trees larger than 14" will be harvested, the logger assuring him that what is left after the harvest will provide perfect conditions for regeneration.

During harvesting, the logger sorts logs on the landing and brokers them to various mills, receiving a premium that amounts to 15 percent to 20 percent more than the stumpage he agreed to pay Uncle John.

1. How should the logger have handled the "tip" from his friend?
2. Does Uncle John have a right to know about the conditions of the referral?
3. How could the logger handle the referral to avoid questionable ethics?
4. Was it acceptable for the logger to visit the woodlot without Uncle John's consent?
5. Given Uncle John's concerns about silviculture (over strictly income production), is it ethical for the logger to act as he did—giving what amounts to silvicultural assurances without the credentials to do so? Is there another way?
6. Is it ethical, in this instance, to buy the timber based on fixed stumpage rates and mill-tallys when the logs were actually brokered from the landing?
7. Are there any conflicts of interest? If so, describe them.

*Questions 1, 2, and 3.* There is nothing wrong with paying a finder's fee for information. If the logger anticipates a business relationship with Uncle John, however, he is obligated to disclose his source, especially if Uncle John asks. The logger might have suggested to the friend that the two of them approach Uncle John together. If the friend is unwilling, the logger should wonder why and consider walking away. Or the logger could approach Uncle John, tell him the source of his information, pay the finder's fee, and risk his friendship with his source. Either way, Uncle John has a right to know that someone gave the logger a tip about his timber. Also—as a friend—the logger may want to advise the tipster to first speak with his wife, since Uncle John is on her side of the family.

*Question 4.* It is common practice in most parts of the country for buyers to "road-cruise" timber and to walk into lots that are not posted. In this case, the logger was doing so based on a tip, and contemplating a future business relationship with Uncle John, he should have first asked permission to walk on his land.

*Question 5.* Unless the logger has credentials to do so, it is unethical for him to advise Uncle John on silviculture. Proper credentials to give that type of advice would include a B.S. in forestry and sufficient experience with similar prescriptions in the timber type. The logger may have many years of valuable experience as a timber harvester, but that does not qualify him as a silviculturalist or as a forester. Also, there is no such thing as a "selective harvest," which presumably means only that someone has selected the trees for harvest. Diameter-limit cutting in this type of circumstance—to regenerate northern hardwoods—is probably the worst possible practice to use. The logger should have sought (with Uncle John's permission) the advice and guidance of a local consulting forester, with the understanding that the consultant would be working for the logger, not for Uncle John. It would be totally unethical for the consultant to approach Uncle John independent of the logger or try in any way to steal the logger's client, even if the consultant believed Uncle John and his forest would be better off not dealing with the logger. The logger runs the risk of Uncle John hiring a consultant of his own, but there is no other way. The logger can only hope that his honesty and forthrightness will eventually be rewarded. If Uncle John is overcome by greed, though, the best thing may be for the logger to walk away.

*Questions 6 and 7.* If it is clearly understood and agreed to by

Uncle John that he is to be paid a fixed rate for stumpage by the logger (who is also the buyer—and this is a key point), then there is nothing wrong with the logger brokering logs to higher-paying markets, assuming Uncle John is paid according to the agreed-upon terms. It looks like a related payment from more than one source, and it would be if the logger's agreement is to pay a percentage of the payments he receives from log buyers to Uncle John. However, assuming the logger is the buyer (which is another reason he cannot honestly advise the client on silviculture without a conflict of interest), there is no problem brokering logs to other markets. Because the brokering is taking place on the landing—that is, on Uncle John's land—the logger has a duty to tell him of this activity and to seek the landowner's approval to do so.

## Final Thoughts

In most parts of the country, the people who make up the local forestry community are well known. Their reputations precede them. One need only listen to hear stories of the thieves and incompetents. The stories of good deeds are not nearly as well known. Nothing will sully a professional's reputation more quickly than dishonesty, and it is almost impossible to reverse a tarnished reputation.

To avoid becoming one more victim, always ask for credentials and references of the people with whom you work. What sounds like a good deal offered from a person of questionable ethics may turn out to be the worst deal of your life. And if you willingly tolerate the questionable practices of others for your gain, you are just as responsible as the person acting in your behalf. Also, expect to be victimized by others if your standards are low.

Remember, if the deal sounds too good to be true, it probably is.

# Chapter 8
# Forest Taxation

"The only certainties in life," according to Benjamin Franklin, "are death and taxes." It is human nature to avoid the inevitability of both, but only one is a matter of degree. It stands to reason then that the objective of a taxpayer is to pay as little as possible, but no less than what is owed. The facts are, however, that most people pay too much, and this is especially true of forest owners. The purpose of this chapter is to describe the three primary forms of taxation that affect woodland owners—property tax, income tax, and estate tax—and to suggest ways forest owners can avoid paying more than is due.

In society, taxes are a necessary hardship, a burden we all share because taxation is one of the only means of raising revenue to expend for the public good. Most people oppose taxation not because of an unwillingness to pay their share but on the definition of "public good." We resent paying taxes because we are unsure how those tax dollars will be used and who will benefit.

From society's perspective, a tax must meet certain criteria to be successful. It must be perceived as reasonably fair and equitable. If it is not, compliance is low and the cost of collection is high. People must also accept the authority or purpose for which taxes are levied, and they must feel that they have a say in how revenue is spent. In the United States, most people believe that taxation should be progressive: those who have more are taxed at a higher rate than people who have less. Of the three types of taxes discussed in this chapter, income taxes and estate taxes are progressive, while property tax is not. Property tax is an example of a flat tax because it is the same rate for all taxpayers regardless of their ability to pay. Sales tax is an example of a flat tax on consumption. Flat taxes are often mistakenly referred to as regressive taxes (i.e., tax impact goes down as income goes up), because the proportion of tax to income is higher for a person of low income than for a person of high income. Tax liability should be simple to calculate, definite, and efficient to administer. People need to know a tax is a regular and periodic obligation, not a whim of the taxing authority. Also, the cost of collecting the tax must be substantially less than the revenue raised—the tax must be efficient. Finally, from society's perspective, a tax must be economically neutral. A tax burden should not cause wild, destabilizing swings in economic activity that would tend to destabilize local economies (Gregory 1972).

Tax equity is an elusive concept. Theoretically, it is the point where one's ability to pay is roughly equal to the taxpayer's perception of benefits received. When ability to pay far exceeds the perception of benefits, the taxpayer is apt to feel as though the burden is excessive. Even if a person has an infinite ability to pay taxes, he is restrained by feelings that his tax burden is out of proportion to the benefits he can expect to receive.

Equity is almost always an issue among forest owners, because land values on which taxes are levied are not necessarily a reflection of ability to pay, and forest lands require few or no services of the community. Because forests do not cost the community very much (nor do they put more kids in the local schools), why then, a forest owner argues, are my taxes so high?

A taxing authority, whether a local town or county, or the federal government, is guided by two principles: (1) It must raise sufficient revenue to cover its obligations, and (2) its policies must promote (or at least not disturb) economic stability while fostering growth at a sustained and acceptable level (Gregory 1972). Local communities traditionally raise revenue through property taxes, while state and federal governments have relied on more progressive taxes, such as the income tax.

## Property Tax

Property tax, also known as an *ad valorem* tax, is levied on the fair market value of real estate and, in some parts of the country, personal property—both tangible property, such as cars and boats, and intangible property, such as stocks and bonds. It is the primary source of revenue for local municipalities, and for that reason it is jealously guarded as one of the few means of local control left to residents.

Local tax rates are determined by allocating the total cost of running the municipality (including the cost of public schools in most areas, which is 70 to 80 percent of the total) over the assessed value of the property in the town, also known as the "grand list." Periodic assessments, usually based on arbitrary rules and often wildly inexact, are used to compile the grand list. When the municipality approves a budget, usually by popular vote, the tax rate is determined by dividing the budget by the grand list. For instance, a town with a $385 million grand list and a budget of $7.5 million will have a tax rate of $1.95 per $100 of assessed value ($7.5MM/$385MM). In some areas of the country, the tax rate is expressed as dollars per thousand dollars assessed value, also known as the millage rate or mil rate. In the example above, the mil rate would be $19.50.

In many areas of the United States, property taxes are the principal source of funding for schools, and the majority of revenue raised goes to the public school system. Wealthy communities have the best schools, but because recent state supreme court decisions (Vermont, New Hampshire, and elsewhere) have found that adequate funding of public education is a state responsibility, the structure of local property taxes—and school funding—is changing. Wealthy communities are enraged about sharing their ability to raise revenue with poorer communities for what they see is a responsibility of state government. Exactly how sharing occurs is still the subject of much debate.

Collecting property tax is a relatively simple matter of multiplying the tax rate times the assessed value of each taxpayer's property and sending out bills. In many communities, taxes are spread over two to four payments in a year. Some communities will offer an incentive rebate to people who pay their taxes in one payment. If a taxpayer disagrees with his assessment, he or she can launch an appeal first with a local board of civil authority, then through the court system. Local appeals are fairly common, but court battles over property taxes are not.

It is pointless to appeal an assessment unless you have documented an error or oversight in the assessor's figures. Unfortunately, this almost always involves neighbors, since their assessments are the basis for comparison. Winning a small concession on your property taxes may not be worth alienating neighbors and friends. Also, launching an appeal on the grounds that all forest land is overtaxed will not work. The review board is charged with reviewing each case independent of another. Its job is to resolve disputes, not to change public policy. Most taxpayers use moral outrage as the impetus for an appeal. They soon learn how quickly moral outrage turns to frustration in front of a citizen's board that must make decisions on the basis of facts, not feelings.

It is common practice, where state law allows, for a community to manipulate assessments so that tax rates appear to be lower than they really are. In this way, a community can attract businesses and development and then shift the tax burden onto the new properties. Some states require municipalities to assess property at 100 percent of fair market value. This is almost always the case in states where there is an attempt to equalize revenue for schools in property-poor versus property-rich towns. Forest owners with land in communities that assess property at less than fair market value have little grounds for appealing their assessments. In states where full fair market value assessments are required by law, forest owners quickly learn that careful forest-management practices and a commitment to long-term forest use have virtually nothing to do with assessment of fair market value.

Property assessment is purportedly based on fair market value of an asset at its potential highest and best use—in an economic sense. The problem with this concept as it relates to forest land is that it requires land be assessed as though it were being held in inventory for development, since development (from the perspective of the keepers of the grand list) is the highest and best economic use of the land. Many tragic stories are told of tree farmers with developable land being taxed out of business and forced to sell. The irony of this situation is that most people in the community would prefer to see forest land remain intact. Studies have also shown that the increased tax revenue from developed forest land usually falls far short of the extra costs to the community for services demanded by people moving into the town, including the cost of new schools for their children. Developed land costs a community more than undeveloped land, and the community that encourages development gets sucked into a revenue spiral that is nearly impossible to escape.

Most states have programs that allow forest land to be taxed at its current-use value rather than fair market value. Requirements vary, but generally the owner must agree not to develop the land. If the land is sold, the new owner usually has the option of continuing the use of the former owner or changing the use and paying a penalty. In some states, a forest owner must grant a lien or an easement—in favor of the state—which stays with the title forever, or until the land is developed or the lien is reacquired by the current owner. The purpose of these programs is to allow individuals to practice long-term forest management without the annual burden of property taxes based on nonforest uses.

Many forest owners do not participate in these use-tax assessment programs because they feel as though the state is dictating how land is to be managed, or they are concerned about clouding the title with liens and easements. Those are valid concerns, but until all property is assessed based on its current—not potential—use, forest land in developing areas will continue to be overtaxed. An owner who intends to keep land intact and to pass on a forest management legacy has more to gain from current-use taxation programs than he or she—or future owners—stands to lose. For more information on these programs, contact your state extension forester (Appendix A) or call the county or town clerk.

Finally, property taxes are considered an annual expense by the Internal Revenue Service and can be deducted each year on Schedule A for people whom the IRS calls timber investors, or on Schedule C or F for people in the business of managing forest land. The difference between "investors" and people who the IRS say are "in the business" is discussed later in the section on reporting income and expenses from forest management. For those investors who do not file Schedule A, annual property tax payments can be added to the cost basis of the property and recovered when property is sold. Adding periodic costs to the current cost basis of property is known as *capitalization.*

### *Yield Tax*

Some states tax the income from timber sales, either in addition to a property tax based on less than fair market value or as a stand-alone tax. When assessed in addition to a property tax, the yield tax is sometimes called a "severance tax." The good thing about a yield tax is that it is assessed when the property is generating income. By itself, a yield tax eliminates the annual economic burden of property taxes (or lessens the burden) in years when there is no income from timber sales. With a yield tax, the owner is given an incentive to grow higher-value products over longer rotations.

Although the yield tax appears to be the perfect solution to increasingly burdensome *ad valorem* property taxes, it has not been widely accepted. In states where it is an alternative to annual property taxes, and taxpayers have a choice, forest owners tend to stick with the property tax. Why? Because they fail to see any clear advantage of one form of taxation over another, and they fear the uncertainty of delaying tax liability to some indefinite time in the future. People would rather pay an annoying but certain amount each year than an unknown sum in the future.

Contact your state extension forester to see if there is a yield or severance tax in your state. Bear in mind that from an economic perspective, a yield tax is much preferable to an annual property tax. Also, in some states it is customary for the timber buyer to pay the tax, or it is a fairly easy matter to set up the sale contract in such a way that the buyer is responsible for it.

## Income Tax

Internal Revenue Service rules as they apply to timber are obscure and confusing. To make matters worse, it is almost impossible to obtain a consistent opinion from the IRS on a timber tax question, and tax laws are constantly changing. What references to timber that do exist in IRS literature are, at best, misleading and in some instances wrong. Even people who are very familiar with IRS rules are lost when it comes to reporting income and expenses of a timber sale, and the experts on timber taxation do not always agree. Why all the confusion? In part, it is because subsections of the IRS Code deal specifically with timber. Those subsections are intended to apply mostly to timber-using companies that also own and manage forests. The rules (Subsections 631[a] and [b] of the IRS Code) allow those companies to treat the sale or cutting of timber as a long-term capital gain. This is a special rule, because timber can be viewed as a wood-using company's inventory, or "stock in trade." Generally, other types of companies—from manufacturers of computer chips to disposable diapers—cannot treat income from the sale or use of inventory (i.e., microchips or diapers) as a capital gain. However, non–wood-using companies can treat the sale of standing timber they might own as a long-term capital gain under the same rules available to wood-using industries. The advantages of the rules to the timber industry (and other business entities) make interpretation of the same rules as they apply to nonindustrial forest owners (nonbusiness entities) confusing and difficult, both for taxpayers and for the IRS.

The purpose of this discussion of income tax rules as they apply to timber transactions is to describe a fairly simple and straightforward way for most woodland owners to report income and expenses. Although the methods presented here meet the information requirements of the IRS under the Paper Work Reduction Act, and have been accepted by at least a few revenue agents in the Northeast, they have not been formally approved by the Internal Revenue Service. Because some of my recommendations, especially as they apply to filing Form T (discussed later), are contrary to strict interpretations of current IRS rules, readers are cautioned to seek other opinions, especially if their circumstances differ substantially from those discussed here.

If you live in a state that has an income tax, chances are the tax is pegged to your federal tax liability, with no special steps necessary to report income from timber sales. In some states, however, the income tax is figured separately, and the rules on capital gains may differ from IRS rules. Also, your state may have a special tax on capital gains that is independent of income from other sources. Special rules may also apply if you live and pay taxes in a state that is different from the state where you sold the timber. Contact your state extension forester for information on local rules that may apply to income from timber sales.

### *Timber Sale Income and Capital Gains*

Internal Revenue Service rules on taxation of income from capital gains changed with the Taxpayer Relief Act of 1997, the first major change in more than ten years. Significantly, the new law is more favorable to investors and so is more forest-owner friendly than before. Both the "holding period" for the sale of assets to be treated as a long-term capital gain, and the tax on gains, have changed, effective on sales after May 7, 1997. The new law increases the holding period from twelve months to eighteen months (except on sales that took place between May 7, 1997, and July 29, 1997, when the holding period is twelve months, even though the new lower tax rates apply). Effective on sales after May 7, 1997, the tax rate on capital gains declines from a maximum of 28 percent (on sales before that date) to 20 percent (on sales after that date) for taxpayers in the 28 percent bracket and higher. For taxpayers who do not exceed the 15 percent bracket, the rate on capital gains is lowered from 15 percent to 10 percent. The eighteen-month holding period applies to all lands acquired before January 1, 2001. These changes are summarized in table 8.1.

For lands acquired in the year 2001 and after, the holding period increases to five years, and the maximum tax rate on capital gains declines further to 18 percent and 8 percent, respectively. Longer holding periods mean less incentive to speculate on investments. In most circumstances, the tax advantages of holding timber for five years will exceed the benefits of quick profits taxed at a considerably higher rate.

Self-employment taxes (a self-employed person's payment into the Social Security system) are not assessed on profits from the sale of capital, whereas "other" income is subject to the tax. Self-employment tax is an important consideration for anyone who is

*Table 8.1. Summary of the Changes in the Holding Period and Tax Rates on Income from Long-Term Capital Gains*

| Date of Capital Gain | Holding Period | Maximum Percentage Tax Rate on Capital Gains Income |
|---|---|---|
| Before May 7, 1997 | One year plus | 28 (or as low as 15) |
| On May 7 but before July 29, 1997 | One year plus | 20 (or as low as 10) |
| On or after July 29, 1997 | 18 months plus | 20 (or as low as 10) |
| On or after July 29, 1997 | One year plus, but less than 18 months | 28 (or as low as 15) |
| Assets acquired on or after January 1, 2001 | Five years plus | 18 (or as low as 8) |

apt to be viewed by the IRS as a sole proprietor or partner in a timber business, or individuals who are retired and have little or no income from wages or other sources. Long-term capital gains income is not taxed for Social Security purposes, and income from capital gains does not reduce Social Security payments. Unless your strategy is to build up your Social Security trust fund account, timber should always be sold in such a way as to allow treatment of profits as capital gains and not as other income. Many inexperienced forest owners report income from timber sales as "other" income on Form 1040 and pay more tax than necessary.

Another reason to report timber sale income as a capital gain: Capital losses can be used to offset up to $3,000 of ordinary income in a particular year, and—a big advantage for some taxpayers—capital losses can fully offset capital gains. This means that a person who, say, loses in the stock market can use those losses to offset the capital gain from a timber sale so that no tax is paid on income from timber. A small consolation, but an advantage nevertheless.

A forest owner should always sell timber in such a way as to allow capital gains treatment (see chapter 6). This means selling standing timber, not logs or the products of trees. It is a fine distinction and one not most IRS revenue agents are apt to question, so even if the method of sale is unclear, a taxpayer should still report income from a timber sale as a capital gain. In most circumstances, capital gains are reported on Schedule D, but to comply with IRS rules, you should attach a page to further explain the transaction. (Attachments to Schedule D are discussed in more detail below.)

### *Establishing Cost Basis of Forest Assets*

One of the major tax advantages of a timber sale is the ability to offset income from the sale with a portion of the original purchase price of land and timber. Known by the unfortunate term depletion, what the woodland owner is doing is subtracting the cost of timber from gross sale income to figure profit. Although used in reference to timber sales, depletion is usually a process whereby a resource owner estimates the extent of a nonrenewable resource, like gravel or coal, then recovers the cost of the resource as it is extracted. The difference is timber volumes and values can be readily estimated and verified, whereas one can only guess the extent of a mineral deposit. When forest owners deplete the cost basis they have tied up in timber, all they are doing is subtracting the original cost of the timber to figure net profit. Depletion is to timber what depreciation is to equipment used in a business. The taxpayer recovers cost as the timber or equipment is used up.

The cost basis of any asset is the sum of its costs up to the present. It is composed of the original cost plus allowable amounts that may have been added to the basis over time, less any amounts that may have been recovered through depreciation or depletion. When an asset is sold, the taxpayer subtracts its current cost basis before figuring gain or loss from the sale.

Although the concept of cost basis is simple, it is not always easy to calculate, especially for timber. Why? Because timber is usually purchased with land, buildings, and other improvements all for one price. Before a woodland owner can recover the cost basis of timber sold, he or she must first determine what portion of the total cost basis of all assets is attributable to timber. This process is one of making an allocation to different accounts. In the example that follows, the owner has set up two accounts, land and timber. Only that part of the total cost basis that is reasonably and fairly allocated to timber is subject to depletion. Land is never depleted until it is sold outright.

For many woodland owners, the cost basis of land and timber is attributable mostly to the original purchase price of the property. However, any carrying charges, such as taxes, interest, and insurance, that were not taken as a deduction in the year they were incurred can be added to the cost basis.

Consider the following example and assume there have been no additions to the basis and no deductions up to the current year:

> *Ms. K acquires 130 acres of forest land in the spring of 1990. She paid $45,500, or $350 per acre. An inventory of the timber was completed by a forester in the fall of 1998. He estimated the current inventory to be 456.2 mbf (thousand board feet) and 1,593 cords. He also estimated average annual growth rate at 2.2 percent. The forester deducted the equivalent of nine years of forest growth (actually, nine growing seasons) to arrive at an estimate of the volume when the land was acquired: 375.05 mbf of sawtimber and 1,310 cords of fuel-quality wood (figure 8.1).*

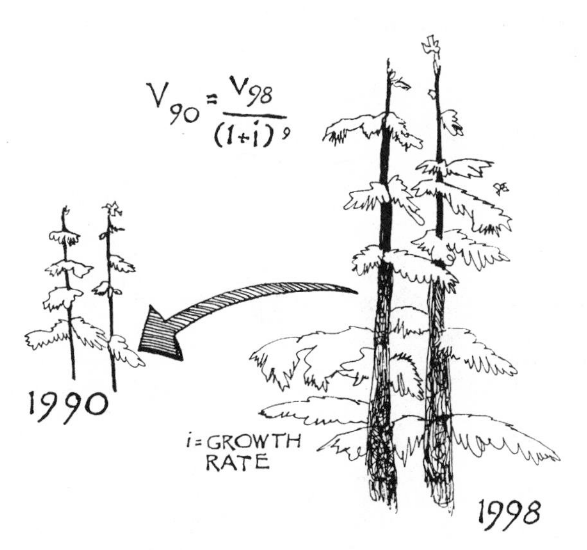


*Figure 8.1. Discounting applied to a current forest inventory is used to estimate inventory at some point in the past.*

The value of the timber inventory on the date of purchase: sawtimber = $31,879 ($85/mbf), cordwood = $9,825 ($7.50/cord). The unit values for sawtimber and cordwood were obtained from information on timber sales in the area in 1990. Total timber value on the date of purchase was $41,704 (sawtimber + cordwood).

The 1990 fair market value of timber is estimated to be about 92 percent of the total purchase price of land and timber combined. Although this may seem reasonable to the woodland owner, it will not seem reasonable to a revenue agent during an audit. *The IRS stipulates that the cost basis of assets must be allocated according to the separate fair market value of each asset independent of the other, and bare land never has a value of zero or less.*

Ms. K investigates the value of her land as bare, cutover property. After much prodding on her part, she finally obtains a realtor's estimate of what her land would have sold for in 1990 without the timber on it. The realtor believes the property might have brought $90 per acre for the tract as bare, cutover land. The fair market value of bare land is $11,700 (130 acres times $90/acre). She has a letter to this effect on the realtor's stationery to document and defend the fair market value of bare land.

With information on the fair market value of each asset (independent of the other), she can allocate her cost basis to the land and timber accounts as shown in table 8.2.

It is important to remember that cost basis is the amount of money that an owner has in the asset, not the current fair market value.

The allocation must be reasonable and fair, not arbitrary. In fact,

*Table 8.2. Allocation of Cost Basis to Land and Timber Accounts*

| Asset | Fair Market Value | % of FMV | Cost Basis |
|---|---|---|---|
| Land | $11,700 | 22 | $10,010 |
| Timber | $41,704 | 78 | $35,490 |
| Total | $53,404 | 100 | $45,500 |

*Note*: The allocation of $35,490 to the timber account is based on Ms. K's calculations that 78% of the total fair market value of land and timber is attributable to timber—$45,500 × 0.78. The balance of purchase price is attributable to bare land—$45,500 × 0.22.

Ms. K may have been able to argue that because of limited access, steep topography, or other constraints on nonforest uses of the property (such as for development), the fair market value of bare land is an even smaller proportion of the total basis of the property. Despite the claims of some economists that cutover forest land can actually have a negative value (after all, who wants to buy cutover land and pay taxes on it?), the IRS insists that some portion of the total cost basis of land and timber be allocated to a bare-land account.

Once the allocation to separate land and timber accounts has been made, the timber account can be further divided into subaccounts for regeneration, young growth, or other product categories. This is especially important if and when Ms. K decides to sell timber. She can create as many subaccounts as she likes, but in this example she sets up two—sawtimber and cordwood. Again, she allocates timber cost basis according to the separate 1990 fair market value of each asset, but divides the cost basis for the asset by its current (1998) inventory to obtain the "unit basis for depletion," as shown in table 8.3.

When timber is sold, its cost basis can be depleted in proportion to the amount of current inventory that is harvested. If 25 percent of the current inventory is harvested, 25 percent of the current timber basis can be recovered. An easy way to do this is to create a unit basis for depletion by dividing the available cost basis by the

*Table 8.3. Allocation of 1998 Timber Cost Basis to Sawtimber and Cordwood Subaccounts*

| Asset | Fair Market Value | % of FMV | Cost Basis (Timber) | Unit Basis for Depletion |
|---|---|---|---|---|
| Sawtimber | $31,879 | 76 | $26,972 | $59.12/mbf |
| Cordwood | $9,825 | 24 | $8,518 | $5.35/cord |
| Total | $41,704 | 100 | $35,490 | |

*Note*: The allocation of timber cost basis to product subaccounts is handled in the same fashion as the original allocation that created the bare land and timber accounts. In this example, it is the separate fair market value of sawtimber and cordwood on the date of acquisition that is used to make the allocation to timber and cordwood subaccounts. The unit basis for depletion is the total in each subaccount divided by the current inventory of the asset. For sawtimber, it is $26,972 divided by 456.2 mbf = the 1998 inventory.

current inventory. In this example, for every thousand board feet Ms. K sells in 1998, she can deduct $59.12 from gross sale income.

In future timber sales, Ms. K must recalculate the unit basis for depletion by completing an inventory before the sale, or by growing the residual inventory by the average annual growth rate (figure 8.2). Whatever basis is available at that time is divided by inventory to calculate the unit basis for depletion. Because forest stands grow (while the cost basis of assets usually stays the same or shrinks as basis is recovered during timber sales), the unit basis for depletion gets smaller and smaller with each succeeding sale. The only time it ever reaches zero is if the entire inventory is sold, or if the land and timber are sold outright.

If you have acquired forest land in the past twenty years, you should investigate the cost basis of your property. Most forest land (especially in the East) acquired before then probably has an original cost basis that is so small compared to the current value of timber as to not warrant the trouble of calculating cost basis. Furthermore, a valuable stand of sawtimber on a good site today may have been a relatively low-valued stand of pole-size timber as little as ten years ago. The allocation of cost basis can be done at any time. However, it must reflect the actual fair market value of timber on the date the property was acquired.

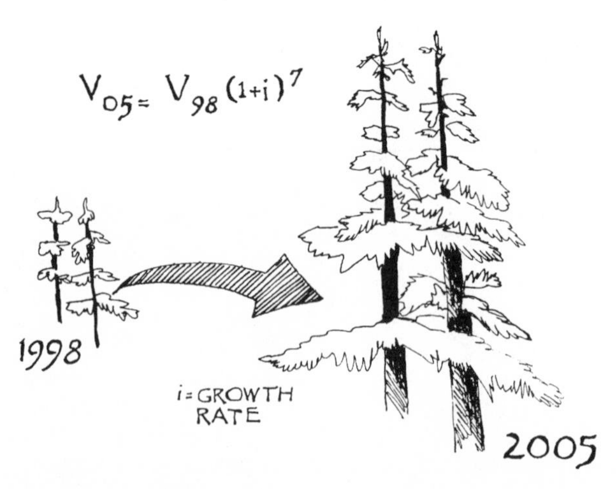


*Figure 8.2. Compounding applied to a current forest inventory is used to estimate inventory at some point in the future.*

When land is acquired by gift, the donor also gives whatever basis he or she has in the land. The cost basis for the new owner is the donor's basis plus any gift taxes that might have been paid. For land acquired by inheritance, the cost basis of the property is "stepped up" to the fair market value as of the date the decedent passed away. This includes the share attributable to a "joint tenant with rights of survivorship" or by "a tenant by the entirety" who dies. For instance, when a husband and wife own woodlands as joint tenants and the wife dies, half the basis they jointly held in the timber is stepped up. This is an important point to keep in mind when planning your estate. Giving land with little or no cost basis may, in the long run, be more expensive to your children than if they were to inherit the land. Also, if you are faced with having to sell timber to help pay the expenses of an illness preceding the death of a spouse, you are better off waiting until after the spouse dies to take advantage of the stepped-up basis.

The 1997 Tax Law also provides changes in how lands in a decedent's estate are valued for tax purposes. (This is discussed in more detail under Estate Taxation.)

## *Forest-Management Expenses*

There are three types of nontimber-sale–related costs borne by forest owners: annual expenses, carrying charges, and capital expenditures.

Keeping track of costs is important because eventually the cost can be used to offset income. This is known as *recovering* the cost, since the effect is to lower taxable income in the year the costs are claimed. The costs associated with timber sales are handled differently. All reasonable and necessary timber sale expenses are deducted from the gross proceeds of the sale to figure capital gain. (Timber sale expenses are discussed in more detail in the next section.)

Annual expenses include such costs as foresters' fees, expenses of maintaining boundaries, equipment depreciation, and other similar expenses that are annually recurring and short-term in nature. Carrying charges include such costs as taxes, interest expenses, and insurance. Capital expenditures include the cost of purchasing more land, permanent improvements to the land (such as a pond or a house), and the cost of planting trees.

In most circumstances, a taxpayer is better off (financially) to recover expenses in the same year they are incurred, because inflation dictates "A dollar today is worth more than a dollar next year." Unless a forest owner is materially and actively involved in managing the land, however, annual expenses can be recovered only to the extent of investment income. Carrying charges can be annually expensed or added to the cost basis of assets, at the discretion of the taxpayer, but capital expenditures must be capitalized and recovered only when the capital is sold (there are exceptions to this rule, as described below). For an investor in timber, there are limitations on the extent to which management expenses can be deducted from other (noninvestment) sources of income.

If a forest owner's annual expenses are substantial, it is an advantage to be "materially and actively" (IRS language) involved in management of the forest. This is the Internal Revenue Service's test of who is and who is not "in the business." Also known as the Passive Activity Loss Rules promulgated in the Tax Reform Act of 1986, the purpose is to severely limit the extent to which limited partnerships can create business losses on paper solely for the purpose of sheltering from taxation income from other sources. Fortunately, for forest owners the test to determine who is and who is not in the business is based on the facts and circumstances of each situation. If a woodland owner (and spouse together) can demonstrate they have contributed at least one hundred hours in a year to the management activities of the forest for which they are claiming the expenses, they will generally be viewed by the IRS as "in the business." This type of taxpayer would show annual forest-management costs on Schedule C or Schedule F. Although Schedule F is for farming businesses and Schedule C is for any type of business, it does not matter which schedule is used, as long as it is used consistently year after year.

If a woodland owner investor does not itemize deductions on Schedule A, there is no way to report annual forest-management expenses and carrying charges, such as taxes. A strict interpretation of IRS rules is that annual expenses not recovered in the current tax year should not be added to the basis of assets. Many taxpayers, however, do so, using the argument that an annual expense not claimed in the current year becomes a carrying charge.

Of the cost items most owners will want to expense—or deduct from current income—in a given year, property taxes should be at the top of the list. Even timber investors who itemize deductions using Schedule A can report property taxes and are not limited by the Passive Activity Loss Rules. However, if you have never reported property taxes on Schedule A (because you have never itemized deductions), you can sum them up for the years you owned the property and add them to the total cost basis. A portion of those costs can then be recovered each time you sell timber.

A woodland owner who is "in the business" can report annual expenses on Schedule C or Schedule F, but that will limit the owner's options when timber is sold. Discussed in more detail below, essentially the caveat is this: If you want to look like a business to take advantage of IRS rules, you must act like a business even in the face of greater financial risks.

Woodland owners who are not in the business of growing forests should try to associate expenses with timber sales. As long as the expense is "ordinary and necessary" (IRS language) it can be used to offset income from the sale of timber.

## *Reforestation Costs*

Costs associated with getting a new crop of trees started are known as *reforestation costs*. These costs must be capitalized; that is, they are added to the cost basis of property and recovered when the property is sold. There is an exception to this known as the Reforestation Tax Incentives. The law allows reforestation costs to be amortized and recovered over the first seven years of the planting. In addition, the incentive includes a 10 percent tax credit, up to a maximum of $10,000 per year, on reforestation costs. A tax credit is a dollar-for-dollar savings in tax liability.

The Reforestation Tax Incentives are a huge advantage in areas of the country where forest stands are replanted after harvest. It allows the taxpayer to recover a major portion of the cost during the first few years of the new stand. Without this program, the taxpayer would need to wait until the first timber sale to recover a portion of reforestation costs borne many years earlier. These incentives are available only for trees that are being planted for fiber production. They are not available to nurserymen and Christmas tree growers.

Also, the tax benefits are subject to recapture if the land is sold for, or converted to, other nonforest uses.

Virtually every state that relies heavily on replanting has detailed information on using the tax credit, which is claimed by filing IRS Form 3468 for individuals (Form 3800 for businesses). The amortization is calculated on Form 4562, which is included in the standard Form 1040 package sent to most taxpayers. The costs of tree planting and site preparation for natural and artificial regeneration are practices eligible for the investment credit and amortization.

For more information on reforestation and related tax incentives, contact your state extension forester.

## *Cost Share Payments*

The U.S. Department of Agriculture has several programs that will provide payments to forest owners who complete certain practices on their lands. The payments are intended to make forest investments more appealing by sharing the costs of such practices as timber stand improvement, tree planting, gaining access, improving roads, and protecting sites from erosion. Generally, these payments may be excluded from income by the taxpayer, but that may not be the best strategy. If payments are excluded, the tax benefit is subject to recapture if the owner does not maintain the practice, or if the use of the land changes. In many instances, though, forest owners receive many thousands of dollars each year, and it may be to their advantage to exclude the payments from income, especially if they have every intention of maintaining the practice. Federal programs that are eligible for exclusion are the Stewardship Incentive Program (SIP) and the Forestry Incentive Program (FIP). The Environmental Quality Incentive Program (EQIP) and the Conservation Reserve Program (CRP) payments currently are not eligible for exclusion from income.

## *Reporting Income and Expenses from a Timber Sale*

It is a wise forest owner, regardless of status as investor or business, who plans forest-management expenses to coincide with timber sales. There is income from the forest to pay expenses, and all ordinary and necessary expenses can be subtracted from the gross proceeds of the sale to figure gain. The more expenses, the lower the gain, the less tax paid.

What are "ordinary and necessary" expenses? Just about any expense that is related to ensuring the timber sale is executed according to plan. Travel, phone calls, lawyers, accountants, and consulting foresters, possibly even a portion of the cost of doing a property survey, are apt to be viewed by the IRS as ordinary and necessary. Overnight accommodations in January at a ski resort for a nonresident woodland owner, including lift tickets (for you and your forester), probably would not qualify as ordinary expenses of a timber sale. To meet the test of "ordinary and necessary," the expenses must first be "reasonable and fair" (IRS language).

A forest inventory and management plan—at least for the areas to be harvested—are ordinary and necessary expenses, as are the costs of gaining access. With a little extra planning, a timber access road can make a fine trail for hiking, skiing, or snowmobiling. A log landing can be cleared with a future house site in mind, and the expense of wildlife habitat improvement—again, with a little extra planning—can be billed as ordinary and necessary timber sale expenses. Don't get too carried away, though. IRS rules say only the portion of a cost associated with the timber sale can be deducted. If you obtain extra services that are too obvious, such as site clearing or trail building, you must estimate the cost of those services and report them as income from the sale.

The profit, or capital gain, from a timber sale is gross income, less expenses, less depletion. If you are an investor in timber (which, the IRS says, most forest owners are), you report the sale on Schedule D and use an attachment to explain sales price and how you arrived at the gain you are reporting (see figures 8.3 and 8.4, later in this chapter). If you are in the business of growing forests, the sale is reported on Form 4797 (with the same attachment) and the result is carried over to Schedule D. Use an attachment to show how you arrived at the sale price and how you have figured gain. Any portion of the sale that results in ordinary gain is reported on Schedule C or F. Readers should feel free to use the wording in figure 8.4 to help draft an attachment that fits their own circumstances.

*Form T.* It is probably not necessary to file IRS Form T, the Forest Industries Schedule, but this is where my opinion differs from other timber tax experts. The infamous Form-T-for-Timber is an information form only—it does not materially alter the tax you owe. Nor does it contain computations you need to perform to ensure your figures are correct. The purpose of Form T is to help the taxpayer calculate and maintain the unit basis for depletion. It is also intended to help the IRS develop a history of the basis of forest assets in the event of an audit. IRS rules say anyone who claims an "allowance for timber depletion" must file Form T. But because the form does not materially alter tax liability, for most forest owners it need not be filed. A taxpayer who can demonstrate acceptable procedures for figuring the cost basis of timber, and can document adjustments to cost basis as assets are acquired or sold, has fulfilled the information requirements of the IRS. However, Form T is an excellent worksheet for figuring unit basis for depletion and keeping track of the basis of assets. For a forest owner who is actively involved in timber sales, or in the acquisition and sale of forest assets, it is good idea to obtain a copy of Form T and keep it up-to-date. But it is not necessary to file it every time timber is sold. An attachment to Schedule D or Form 4797, typewritten on a plain sheet of paper and stapled to the appropriate form, should suffice to explain the particulars of a timber sale with enough detail to allow the IRS to do its business if your return is questioned.

To obtain a copy of the correct Form T, be sure you specify the "Forest Industries Schedule." Another Form T is used to report income and expenses of a trust; if you ask for "Form T," more than likely you will get the trust document. Tax forms are now available on-line at www.irs.ustreas.gov.

A strict interpretation of IRS rules states that anyone who claims a deduction for depletion of timber assets must file Form T. Some timber tax experts believe this is especially true for taxpayers who are claiming to be a business. Those same experts believe Form T filed in the year of a timber sale will help prevent an audit, although there are no data to support this claim. Most revenue agents are not familiar with Form T, so seeing it attached to a return that is being examined may imply the taxpayer knows what he is doing. Regardless, there should be no penalty for not filing Form T if you have handled the cost basis of timber assets properly and the correct amount of tax has been paid. A properly prepared attachment to Schedule D should suffice, especially for timber investors—the vast majority of people who sell timber.

*A Timber Sale Scenario.* If you have never reported income from a timber sale, chances are good that much of what you have read in the past few sections is vague. The purpose of this section is to summarize procedures for accurately reporting income from a timber sale, and to give an example of how it is done.

You need three pieces of information to report the income and expenses of a timber sale: (1) preharvest inventory and average growth rates, (2) an accurate estimate of harvest volumes, and (3) a composite summary of timber sale income and expenses.

If this is your first sale and you have not yet established the cost basis of forest assets, you will also need historical information on stumpage and bare land prices in your area. At a minimum you will need to know how much your timber was worth (fair market value) on the date you acquired the land, and how much the bare land was worth independent of the timber.

It is almost impossible for most forest owners to determine items 1 and 2 without the assistance of a forester. Item 1 is necessary to determine the unit basis for depletion preceding a sale. And, if you have not yet allocated the cost of land and timber to separate accounts, average growth rate is used to grow the stands in reverse, back to the date you acquired the land. Known as discounting, the concept is exactly the same as compounding the interest and principal in a savings account, but in the other direction (see figures 8.1 and 8.2). A caveat of discounting is that it assumes average annual growth rate is constant, which is not usually the situation in forest stands. So long as the forests in question have not changed product classes (from cordwood-size trees to sawtimber-size trees) during the discounting period, however, using an average annual growth rate for periods of ten years or less should not be a problem. Reinventory stands after ten years.

The second item—an accurate estimate of harvest volumes—is necessary for bookkeeping purposes. By subtracting actual harvest volumes from presale inventory, you can estimate the postsale inventory. By using average annual growth, you can grow postsale inventories to estimate available volumes for another sale within a ten-year period. For instance, if you have a sale in the same area three years after the first sale, just grow the inventory left after the first sale by the average annual growth rate (see figure 8.2). This will tell you the current volume immediately preceding the sale, which is used to calculate the unit basis for depletion for that sale.

The third item—a composite summary of timber sale income and expenses—is necessary for obvious reasons, but the information is not always readily available. If you work with a consulting forester, part of his or her services should be to maintain a summary of timber sale income for you and to deliver a copy when the sale is completed. If you work directly with a timber buyer, it is usually your responsibility to keep track of receipts from the mill. Keeping a record of expenses is also your responsibility. Expense claims of more than $25 will require a receipt, and only ordinary and necessary expenses are allowable. An expense that is only partially attributable to the timber sale must be apportioned accordingly. For instance, if you obtain legal advice on a timber sale contract at the same time you are obtaining legal advice on other matters, only the portion of the cost of services associated with the timber sale can be counted.

Consider the following scenario. It is a continuation of the cost basis scenario described earlier in this chapter. If you are not familiar with the case of "Ms. K" and her forest land, you may want to first review the facts presented earlier.

Ms. K decides to sells timber in 1998. Her forester has completed a presale inventory and discounted the current inventory for nine growing seasons at an average annual growth rate of 2.2 percent. With these data, and with information on timber prices and cutover forest land prices in 1990, he helps Ms. K allocate the original purchase price to separate land and timber accounts. He takes the cost basis attributable to timber and figures the unit basis for depletion available in 1998.

In the fall of 1998, Ms. K and her forester implement a "timber stand improvement" prescription across the entire property that yields 149.5 mbf of low-quality hardwood sawtimber and 455 cords of fuelwood. They used a "lump-sum" timber sale method (discussed in chapter 6), and received $85 per thousand board feet of sawtimber and $10 per cord of fuelwood. The other facts of the sale are as follows:

Gross Income $17,257.50

Sale Expenses:
$1,800.00 Forester fees
80.00 Phone calls
175.30 Travel expenses
250.00 Lawyer fees
Total $2,305.30

Depletion Allowance for Timber:
Harvest Volumes x 1998 Unit Basis for Depletion = Depletion Allowance
149.5 mbf x $59.12/mbf = $8,838.44
455 cords x $5.35/cd = $2,434.25
TOTAL $11,272.69

Figure Long-Term Capital Gain:
Gross Income - Sale Expenses - Depletion Allowance = LTCG
$17,257.50 - $2,305.30 - $11,272.69 = 3,679.51

Because Ms. K is apt to be viewed by the IRS as an investor in timber, she can use the lump-sum sale method. If Ms. K were apt to be viewed as in the business of growing forests (or it was her intention to appear to be so), the lump-sum method is not available to her. She must use a sale method where she is at risk up until the timber is measured and paid for. In IRS jargon, this is known as "retaining an economic interest" in the timber. Forest owners who are (or appear to the IRS to be) in the business must file under the provisions of either IRS Code Section 631(a)—if the taxpayer cuts the timber—or under Section 631(b)—if standing timber is sold to a buyer.

A 631(b) sale can be designed to afford the same protections of a lump-sum sale by adding two conditions to the contract: (1) Specify the seller retains ownership of the timber up until the timber is cut, and (2) create a provision for adjustments at the end of the sale to account for a higher or lower actual harvest volume. These two elements fulfill the "retained economic interest" and "uncertainty" requirements of a 631(b) sale but create a contract that looks like (and affords the protection to the woodland owner of) a lump-sum sale.

For more information on subsections 631(a) and 631(b), see *Forest Owner's Guide to the Federal Income Tax* (Siegel et al. 1995).

As an "investor," Ms. K reports the sale on Schedule D with an attachment that explains how the capital gain is calculated (figures 8.3 and 8.4).

*Deductions for Timber Losses.* Losses in timber caused by fire, storm, flood, or any other sudden, unpredictable event can generally be deducted against income in the year the loss occurs. The loss must be "sudden, unexpected, and unusual" (IRS language) to qualify as a casualty loss. The IRS has taken the position that losses from drought do not meet the test. However, there are

**Part II Long-Term Capital Gains and Losses—Assets Held More Than One Year**

| | (a) Description of property (Example: 100 sh. XYZ Co.) | (b) Date acquired (Mo., day, yr.) | (c) Date sold (Mo., day, yr.) | (d) Sales price (see page D-3) | (e) Cost or other basis (see page D-4) | (f) GAIN or (LOSS) FOR ENTIRE YEAR. Subtract (e) from (d) | (g) 28% RATE GAIN * or (LOSS) (see instr. below) |
|---|---|---|---|---|---|---|---|
| 8 | TIMBER AND | 4-1-90 | 10-15-98 | 14,952 20 | 11,272 69 | 3,679 51 | |
| | CORDWOOD | | — | SEE ATTACHMENT — | | | |
| | | | | | | | |
| | | | | | | | |

| Line | Description | | Amount (d) | (f) | (g) |
|---|---|---|---|---|---|
| 9 | Enter your long-term totals, if any, from Schedule D-1, line 9 . . . . . . . . | 9 | | | |
| 10 | **Total long-term sales price amounts.** Add column (d) of lines 8 and 9 . . . | 10 | 14,952 20 | | |
| 11 | Gain from Form 4797, Part I; long-term gain from Forms 2119, 2439, and 6252; and long-term gain or (loss) from Forms 4684, 6781, and 8824 . . | 11 | | | |
| 12 | Net long-term gain or (loss) from partnerships, S corporations, estates, and trusts from Schedule(s) K-1 . . . . . . . . . . . . . . . | 12 | | | |
| 13 | Capital gain distributions . . . . . . . . . . . . . . . . | 13 | | | |
| 14 | Long-term capital loss carryover. Enter in both columns (f) and (g) the amount, if any, from line 14 of your 1996 Capital Loss Carryover Worksheet . . . | 14 | | ( ) | ( ) |
| 15 | Combine lines 8 through 14 in column (g) . . . . . . . . . . . . | 15 | | | |
| 16 | **Net long-term capital gain or (loss).** Combine lines 8 through 14 in column (f). . . . . . . . . . . . . . . . . ▶ | 16 | | 3,679 51 | |

***28% Rate Gain or Loss** includes all gains and losses in Part II, column (f) from sales, exchanges, or conversions (including installment payments received) **either:** • **Before** May 7, 1997, **or**
• **After** July 28, 1997, for assets held more than 1 year but **not** more than 18 months.
It also includes **ALL** "collectibles gains and losses" (as defined on page D-4).

*Figure 8.3. Sample Part II, Schedule D, Form 1040, reporting timber sale income for 1998. Note the reference to an attachment that explains how sale price is calculated.*

**Attachment to Schedule D (Form 1040)**
**1998**
**for**
**Mr. K.**
**123-45-6789**

**Volume of timber sold**:

*149.5 Mbf of mixed hardwoods and 455 Cords of fuelwood in standing culls and tops.*

**This was a lump sum sale that took place during October and November of 1998. Timber was paid for at the following rates specified in our contract:**

*Sawtimber:* *$85.00/ Mbf*
*Fuelwood:* *$10.00/Cd*

*Recovery of the cost basis in timber through depletion is based upon an allocation of the original purchase price of land and timber to "Land" and "Timber" accounts according to the separate fair market value of each asset on the date of acquisition (4-1-90). The following unit rates of depletion are based on the total adjusted cost basis in the timber account divided by the standing inventory of timber and fuelwood immediately preceding the sale:*

**1998 Depletion Units**

**Sawtimber:$59.12**
**Fuelwood: $5.35**

**Sale price (col. d) is calculated as follows:**

*Gross Income* *$17,257.50*

*Less the following sale related expenses:*

| | |
|---|---|
| *Consulting forester* | *$1,800.00* |
| *Phone & Travel* | *255.30* |
| *Lawyer Fees* | *250.00* |
| | --------------- |
| *Sale Price:* | *$14,952.20* |

**Cost or other basis (col. e) is calculated as follows:**

*149.5 Mbf X $59.12/Mbf plus 455 Cds X $5.35/Cd = $11,272.69*

**Gain (col. g) is calculated as follows:**

*Sale price ($14,952.20) - Cost basis ($11,272.69) = $3,679.51*

*Figure 8.4. Sample attachment to the Schedule D, Form 1040, in figure 8.3, describing how sale price is determined.*

instances when—on appeal—losses from drought of "unprecedented and extraordinary severity" were treated as casualty losses. Losses from pollution, insect infestations, or disease depredations usually are not considered "sudden, unexpected, or unusual" enough to be accepted as a casualty loss. Also, even though a flood can create a casualty loss, losses from flooding in flood-prone areas do not qualify.

A casualty loss cannot exceed the timber's adjusted basis (this is not the same thing as fair market value). The loss is figured by estimating the gross loss, then subtracting income received from salvage and any insurance payments.

For more information on casualty losses of timber, see *Forest Owner's Guide to the Federal Income Tax* (Siegel et al. 1995).

*Form 1099.* Whenever you receive payments of $300 or more from another nonwage source, the payer is required to file a Form 1099 with the IRS. The purpose is to allow the IRS an easier method of tracking money changing hands. If you worked with a consulting forester during a timber sale, the forester may have acted as the banker for the transaction (assuming you established a fiduciary relationship with the consultant). All stumpage or log payments went into his escrow account and from that account he drew his payment or commission, and then he issued payment(s) to you. He is required to file a Form 1099-S reporting the gross income from the timber sale (less his commission). The figure reported on Form 1099-S is "gross sale income" used to figure "sale price" in Part II, Column d of Schedule D (see figure 8.3). If the forester has already deducted his commission and other expenses of the sale, you must not subtract those sums again from the Form 1099-S amount to figure sale price. You may have other expenses, though, such as phone charges or attorney fees, to subtract from gross sale income.

If you receive stumpage payments directly from the buyer, every time you receive a payment of more than $300, the buyer is required to file a Form 1099-S. This is especially true for Section 631–type sales. The sum of those payments is used to figure the sale price on Schedule D. When you make payments to the consulting forester (if that is the arrangement you have) or to others for services, you are required to file a Form 1099-MISC.

It is important to keep these forms straight. IRS computer programs that scan taxpayer returns look for 1099 income in certain places. The 1099-S is for "Gross proceeds from real estate transactions" (IRS language), so the program "looks" on the taxpayer's Schedule D—just where you want it to look. The 1099-MISC is for nonemployee compensation, which is exactly what your payment to the forester is. The program will look for this income on the forester's Schedule C. Also, don't be surprised if the timber buyers and foresters you work with do not issue 1099s. Even though it is the law, compliance is low. Don't rely on 1099s from people you do business with for record-keeping purposes. It is not unreasonable, though, to request 1099s of the buyer or forester—in writing—before the sale begins.

*Handling an Audit.* Many forest owners believe that reporting the income from a timber sale as a long-term capital gain will trigger an audit. This is not true. Although in a particular year the IRS may screen returns using capital gains or losses and other parameters as flags, Schedule D is used so commonly it is automatically included in the Form 1040 package. Your chances of an audit are no greater than almost anyone else's, probably less than 2 percent in any tax year.

In the unlikely event that your return is examined, keep meticulous notes and records pertaining to the timber sale. Maintain a diary of management activities and the time devoted to each. The IRS is usually three years behind, so the only way to survive an examination is with good records. Another reason for good records: so the examination can be handled by mail. The burden of proof is on the taxpayer not the IRS (although this appears likely to change within the next few years). Even though the revenue agent will come across as an expert on timber sales, he or she probably knows very little about them. Appropriate forms and procedures aside, the agent is looking to see if you have been reasonable and fair. If you have stretched interpretations of IRS rules so they are always in your favor, your return will be questioned.

Always request that an IRS forester is present during the audit (if the main subject of the audit is timber related). The forester, of which there is at least one in each IRS region, will act as an interpreter for the auditor, since IRS auditors know very little about expenses and income from timber sales.

The most expensive aspect of settling a dispute with the IRS that goes against you is the interest penalty on the tax, figured from the date the tax was due. You can either pay the tax and penalty or make a deposit with the IRS until the matter is settled. The IRS will refund excess tax and penalties or deduct from your deposit only the amount you owe. A deposit may be withdrawn anytime during the settlement proceedings, but a disadvantage is that the principal does not accrue interest.

If an examination cannot be settled by mail, request the audit be held on or near your forest land if questions about the return are mostly related to the forest. That way, you can easily visit the property if necessary to help explain a point. The location of the audit is the prerogative of IRS, but inviting the revenue agent to your woodlands is a good way to garner home court advantage and to show exactly what goes on when forest lands are managed.

## Estate Taxation

Some people think one of the consolations of dying is that you do not need to pay taxes anymore. Right? Wrong! When a person dies, the IRS levies an estate tax on all of your assets. In 1997, up to $600,000 could escape estate taxation through a tax credit of $192,800. The credit is subtracted from the total estate tax due, so if the estate was $600,000 or less in 1997, there is no tax. In the others words, up to $600,000 of total estate value was excluded from estate taxation.

The Taxpayer Relief Act of 1997 has increased the exclusion from $600,000 (before the act and in 1997) to $1 million in the year 2006. Table 8.4 shows how the excluded amount will increase. After 2006, the exclusion amount will be adjusted each year using an inflation index. Because one of the chief complaints of forest owners before the 1997 Tax Law was the failure of the Unified Gift and Estate Tax Rules to account for appreciation of assets due solely to inflation, the new rules are welcome relief to forest-owning families. Despite this, the new law just barely covers past effects of inflation on appreciation of assets. Depending on the "inflation index" used, we can only hope the exclusion will keep pace with inflation after the changes are fully implemented.

Under the Unified Gift and Estate Tax rules, a taxpayer is allowed to make an unlimited number of gifts of $10,000 or less per donee each year that are not subject to any transfer tax. These are nontaxable gifts (discussed further in chapter 9). Gifts over this amount will use up a portion of the exclusion in table 8.4. So long as the sum of taxable gifts and total taxable estate does not exceed the values in table 8.3, no estate tax is due when the taxpayer dies. On amounts that exceed table values, the tax rate starts at 37 percent, increasing to a maximum rate of 55 percent. The annual $10,000 gift allowance will increase each year under the new tax law at a rate indexed to the rate of inflation.

A taxpayer who dies in the year 2006 can have made up to $1 million in *taxable* gifts and bequests without being taxed. If the combined value of taxable gifts and the fair market value of his estate exceeds $1 million, the excess is taxed at rates that start at 37 cents on every dollar greater than this amount. There is one exception to this, known as the *marital deduction.* A spouse can leave an estate of unlimited size to an eligible surviving spouse, but when the survivor dies, the Unified Gift and Estate Tax Rules then apply

*Table 8.4. Effect of the 1997 Taxpayer Relief Act on the Unified Gift and Estate Tax Rules*

| Year | Exclusion Amount |
|---|---|
| 1997 | $600,000 |
| 1998 | 625,000 |
| 1999 | 650,000 |
| 2000 | 675,000 |
| 2001 | Unchanged |
| 2002 | 700,000 |
| 2003 | Unchanged |
| 2004 | 850,000 |
| 2005 | 950,000 |
| 2006 | 1,000,000 |
| 2007 | Increased annually according to the rate of inflation |

to the survivor's estate. If the value of the married couple's assets was greater than $625,000 in 1998—which is often the case for people who own even modest-size woodland tracts in many parts of the country—when one spouse dies, the surviving spouse can protect only the first $625,000 from estate tax in 1998. If the surviving spouse lives until 2002, up to $700,000 can be protected. Amounts in excess of this are taxed at the rates listed in table 8.5.

The marital deduction lulls people into thinking there is no need to plan ahead. With the exception of the marital deduction, however, the Unified Gift and Estate Tax Rules apply to individuals. Families that own forest lands are often astounded when the estate of a parent of modest means turns out to be far in excess of expectations. But a deceased "land rich, cash poor" parent leaves

*Table 8.5. Unified Gift and Estate Tax Rates*

| Column A<br>Taxable amount over | Column B<br>Taxable amount not over | Column C<br>Tax on amount in column A | Column D<br>Rate of tax on excess over amount in column A |
|---|---|---|---|
| 0 | $10,000 | 0 | 18% |
| $10,000 | 20,000 | $1,800 | 20 |
| 20,000 | 40,000 | 3,800 | 22 |
| 40,000 | 60,000 | 8,200 | 24 |
| 60,000 | 80,000 | 13,000 | 26 |
| 80,000 | 100,000 | 18,200 | 28 |
| 100,000 | 150,000 | 23,800 | 30 |
| 150,000 | 250,000 | 38,800 | 32 |
| 250,000 | 500,000 | 70,800 | 34 |
| 500,000 | 750,000 | 155,800 | 37 |
| 750,000 | 1,000,000 | 248,000 | 39 |
| 1,000,000 | 1,250,000 | 345,800 | 41 |
| 1,250,000 | 1,500,000 | 448,300 | 43 |
| 1,500,000 | 2,000,000 | 555,800 | 45 |
| 2,000,000 | 2,500,000 | 780,800 | 49 |
| 2,500,000 | 3,000,000 | 1,025,800 | 53 |
| 3,000,000 | — | 1,290,800 | 55 |

*Note*: Figures are the tax rates before application of the exemption, which creates a tax credit of $192,800 in 1997. After the tax credit, the estate is taxed at rates beginning at 37%.
*Source*: "Instruction for IRS Form 706."

children with the problems of paying the estate taxes—both federal and state. The children are forced to sell—usually to the highest bidder—and productive forest land is converted to condominiums. The estate tax is due within nine months after the decedent passed away. The executor of a decedent's estate can elect (for reasonable cause) to make payments in installments over twenty years under the new law, but interest is charged on the balance due during the installment period. The good news is the interest rate has been lowered from 4 percent on the outstanding balance to 2 percent.

The Taxpayer Relief Act also immediately increases the exemption for small businesses and family farms, including tree farms, to $1.3 million. Over the next ten years, the exemption will gradually decline to $300,000. With the large immediate exemption come restrictions, but taxpayers who qualify as "in the business" under existing IRS "material participation" rules generally would qualify for the family-farm exemption.

Another valuable aspect of the 1997 Tax Law as it applies to estates that include forest lands: An executor may elect to exclude up to 40 percent of the value of land from which a conservation easement has been donated to a "qualified charitable organization." The land, though, must be located "within a 25-mile radius of a metropolitan area, a national park or a wilderness area, or within 10 miles of an urban national forest" (IRS language). An exception to this provision is that an estate that takes advantage of the family-farm exemption cannot also take advantage of the easement discount. Because the family-farm exemption is declining, a taxpayer may want to consider strategies that take advantage of the conservation easement, discussed in more detail in the next chapter.

## Chapter 9

# Planning for Woodlands in Your Estate

The most difficult aspect of estate planning is contemplating one's mortality. It is hard enough to effect good planning over just a few years of one's lifetime. How then can one expect to plan for an indefinite future of which he or she is not even a part? Estate planning is tough business, and most people put it off until the last minute, or until it is too late. Mixed feelings about children, an unstable marriage, or a belief the estate is so small as to not warrant planning are other reasons people delay. If you own forest land and it is your intention to keep the land intact and actively managed for woodland resources after you are gone, an estate plan is essential. An individual's estate is the sum of what he or she owns: cash, personal belongings, intellectual property (copyrights, patents, etc.), retirement plans, life insurance, investments, and real estate.

What makes the need for estate planning of woodlands more important than for other assets? Plenty of reasons, not the least of which is that a reasonable planning horizon for most forest types typically exceeds a lifetime. From the perspective of decisions that are strictly good for the forest, this more than any other reason is why forest planning should always look beyond the current owner's tenure (figure 9.1). Another reason, from a strictly financial perspective: Woodlands have appreciated at a rate that far exceeds inflation in many parts of the country. It is not uncommon for a

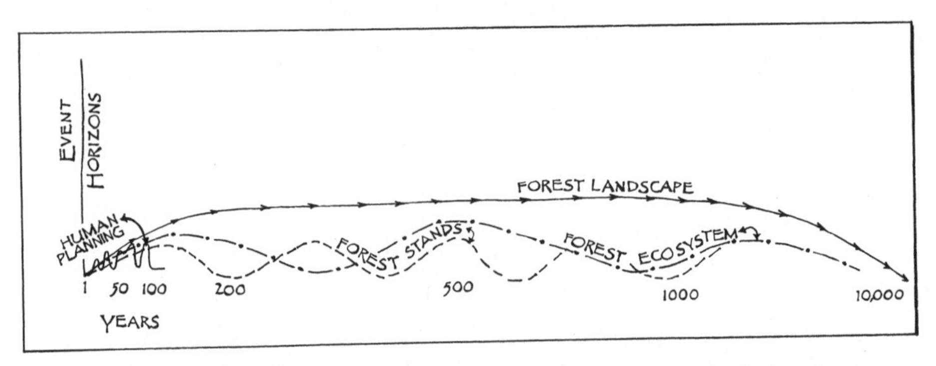


*Figure 9.1. People, forest stands and ecosystems, and whole landscapes exist in different event frequencies. Human circumstances can change month to month or year to year, while ecosystem changes—in the absence of human intervention—take place over a much longer time frame. Good forest-management planning should exceed a single generation and recognize events at least at an ecosystem level.*

family that always thought of itself as "poor" to discover an enormous equity in forest land. These cash-poor families often face hefty estate taxes within nine months of the decedent's death. Often, the only way to settle with the IRS is to sell off land and/or timber.

Estate planning is a primary issue of our times, as baby boomers watch their parents age, and parents wonder, "What do we do with our property?" Much has been written on the subject, and many excellent books cover all the angles. There are even some do-it-yourself books for people who hate to pay for legal advice. Listed at the end of this chapter are references to book titles on estate planning (see References and Suggested Readings) that are readily accessible in local bookstores. Forest owners who have not yet done any estate planning are encouraged to follow up this chapter with one or more other books that delve into the subject in a way that is far beyond the scope of this book. However, as much as I advocate woodland owners being actively involved in managing their lands, estate planning is not a do-it-yourself proposition. To effect a good estate plan, you need solid legal advice from someone who has experience with that body of law. Even a small, seemingly inconsequential mistake in documentation can void the entire effort, leaving the family more confused and distraught than if you had done nothing at all.

People often assume the way to avoid the hassles of estate planning is to leave everything to your spouse. The problem with that strategy is, even though a spouse can inherit free of tax an estate of "unlimited" value, eventually the surviving spouse will die, and IRS rules exempt only the first $600,000 to $1 million, depending on when the surviving spouse dies. The death tax on the first spouse to die is deferred until the surviving spouse dies. One goal of a proper estate plan is to arrange the assets of a family (i.e., a married couple and their heirs) to avoid paying death taxes on amounts up to the maximum exemption of the first spouse to die, as well as on the maximum exemption available to the surviving spouse. If the total estate is worth more than the maximum exemptions available to each spouse, additional strategies can be employed to further minimize estate taxes. Gifts, donations to qualified organizations, and other methods can be used to make the estate tax as low as possible. Finally, a proper estate plan anticipates the taxes and other expenses that might be due when the surviving spouse dies and ensures a ready source of cash to pay the tax and other expenses without having to liquidate assets.

Transfer costs include all the expenses the family will sustain to settle the affairs of the final estate. These include federal and state death taxes and administrative expenses, such as the cost of probate to settle with creditors and disperse the estate to heirs. Probate is the legal process of proving the validity of a will and securing authority from a court to represent the deceased and to carry out the will's provisions. A will is a properly executed legal document in which a person declares to whom his or her possessions are to go after death. When a person dies without a will, the estate is said to be "intestate." Possessions usually go to the surviving spouse and to children based on a formula defined by state law. If there is no surviving spouse and no will, the probate court resolves questions of who gets what, again by formula. Probate can take six months to three years or more, depending on the complexity of the estate and arguments raised by people who make claims against the estate. Probate can be avoided with good estate planning. Because it is a public process, though, probate is an excellent way to keep people honest.

Precisely what is estate planning and what is the connection to forest-management planning? It is a process to consider the value and disposition of all real and personal property and property rights owned by a person or a married couple. It is usually done in anticipation of dispersing property to others, with three things in mind: (1) to continue a forest-management legacy and to keep land intact and in the family; (2) to minimize the cost of transferring ownership when the estate is dispersed; and (3) to provide for dependents and heirs.

Continuing a forest-management legacy means providing guidance to the people who will make management decisions in the future. It is their job to ensure wildlife habitats are protected, roads are maintained, timber grows to rotation, and stands are adequately regenerated. It is your job as planner to provide leadership, direction, and continuity with clear statements about how the land is to be managed. It is easy to do this without being dictatorial. You also want to avoid tying the hands of your heirs. They must be able to react to conditions in the future that you cannot predict. The idea is to keep the land in the family, to inspire your children to work with and enjoy the forest as much as you do, and to give them a course to follow with latitude to make their own decisions. You probably also want to prevent them from drastically changing the use of the land, or selling it to someone who has no intention of managing the forest. This is accomplished by putting covenants into the title, or by separating the bundle of rights through the use of easements (see chapter 3). Many forest owners are reluctant to do this, reasoning the restrictions on use will decrease the marketability of the land. It is true—all other things equal, a buyer will pay less for a title with restrictions than one without. Many experts in this field, however, believe we will see strong markets for protected lands in the future. This means your forest could actually increase in value as a result of restricting its title in your estate, so that if your children decide to sell, they will have no problem locating a buyer who wants to continue managing the woodlands. Also, more often than not the tax benefits of giving an easement for conservation purposes to a local land trust are attractive. (This is discussed in detail in the section on charitable contributions.)

Aside from providing a long-term plan for forest resources, estate planning is a means of providing for dependents and heirs after you pass away. The plan identifies who will make decisions in your absence. It also provides a means for a ready source of cash to pay day-to-day expenses, outstanding medical bills, and any other expenses needed to settle affairs. A fairly common strategy is to purchase life insurance and name a spouse, a trust, or your heirs as beneficiaries. Life insurance is not probated and it is usually paid fairly soon after death. Do not make the mistake of naming your estate as the beneficiary, though, which would make the life insurance benefit part of your probated estate. Your insurance agent can offer more information on life insurance policies and payout options for when you die. Another ready source of cash is a jointly held savings or checking account. If you specify that the co-owner of the account has rights of survivorship, the funds in the account belong to—and are accessible by—the joint owner when you pass away.

It is often a good idea to grant a durable power of attorney (see chapter 6) to a spouse or family member. This will allow someone to act in your behalf if you become incapacitated. A common legal error among spouses is to execute a general power of attorney, which is not recognized as valid after the principal becomes incapacitated, although state laws on this are changing. Even with a durable power of attorney, it is a good idea to provide a copy to people with whom you do business so there is no question if something happens to you. A word of caution: A durable power of attorney is an extremely powerful legal document that authorizes another individual to act in your behalf. You must trust that person implicitly. The document creating a durable power of attorney can be drawn up, signed, and held by your lawyer with specific instructions about the circumstances under which you want to have the document passed to your designee. A durable power of attorney expires when you do, so the person who holds it must be cautioned not to try to use it after you have passed away.

As a woodland owner, it is very important to involve your children in the estate-planning process. To a large extent, the success of your efforts will depend on their interests in following your forest-management plans when your estate is dispersed to them. However, in many instances, the owner assumes a son or a daughter will take over the tree farm, but the children have no interest. This is not a problem as long as you know their feelings before leaving them the land, and they in turn understand and respect your commitment to the woodlands and will honor your plans. Even if your children do not want to be directly involved in the continued management of the family forest, you can still keep the land intact and devise a means for them to benefit from woodland income.

When it comes to estate planning, families do not communicate well. Yet it is extremely important that your heirs understand and accept your plans. Sometimes the best way to bring up the subject of estate planning with children is to ask them outright what they would like to see. You may be surprised to learn they have already given much thought to the final disposition of your assets. This is one of the reasons children do not bring up the subject—it appears as if they are waiting for you to die.

To have a meaningful conversation with children about these matters, it is important to make them feel comfortable so they speak their mind. Do not focus on the dying and loss. Rather, think about what you are leaving behind. Remember, "A whole [life] is that which has beginning, middle, and end" (paraphrased from Aristotle). Dying is as much a part of life as living. To die well is to do so in consideration of the people you leave behind. And it is an enormous responsibility to die without having given careful consideration to the long-term disposition of your forests.

## The Estate-Planning Process

The first step in planning your estate is to learn more about the subject. Many excellent resources are available, and workshops on estate planning for woodland owners are common in all parts of the country. Because you will need to assemble a team of experts to pull the plan together, your job is to become a well-informed consumer. Most of the current literature on estate planning assumes the reason for your interest is to shelter assets from taxation. Although this is a valid concern, there are other reasons for planning. The discussion that follows emphasizes reasons for estate planning that have to do with keeping forest lands in the family and continuing your forest-management strategies.

Once you know a little more about estate planning, generate a wish list of things you would like to see happen after you are gone, just as though you were going to live indefinitely. For instance, there may be a stretch of stream on your land you have always wanted to improve for brook trout habitat. You realize now you will never get to that project in your lifetime, but you still want to see it done. Write it down.

Consider what the future might hold for forests in your area. If you believe surrounding lands may be developed in the future and important habitats destroyed, you may want to dedicate your forests, or a portion of them, to providing critical habitat for species that might be threatened. Write it down.

Perhaps you have a favorite stand of trees that you had hoped would one day become large, very highly valued, veneer-quality logs. Although the stand could be harvested now for a substantial profit, you think it best to wait. Write it down. And what about other favorite places—where you like to hunt, or ski, or picnic, or special places in your woodlands that you find spiritually fulfilling? There may be areas of historical significance that you want to see protected, such as caves or other special habitats—anything you think is important that you would like to see managed, protected, or used after you die, write it down.

Next, assemble a list of your assets. Include everything of value to you and your family, whether it has any monetary value or not. You would be surprised to learn that something you regard as a worthless bauble is a treasure to one of your children. For the assets that have titles, find out how the property is held (see chapter 3). When you hire an estate planning attorney, these are some of the first questions he or she will ask. If you can, try to estimate the current fair market value of each asset. Even a rough estimate will help when it comes time to partition the estate for tax purposes. Estimate the rate at which your estate is appreciating. Because the new tax law changes the maximum exemption amount each year, you need to have some idea how your net worth increases.

Meet with your children. Tell them about your efforts to plan your estate, with special emphasis on the woodlands. The importance of this discussion is to ascertain their interest in carrying on as managers. It will also give you a sense of how to balance the value of assets to proportionally disperse your estate. For instance, if your daughter is interested in the forest (even without the development rights), you may want to offset this with a bequest of stock, bonds, or other assets to her siblings. Also, if your children express interest in specific items, such as antiques, furniture, art, or other objects, you may want to affix a note or label to the piece with the child's name on it, besides making a bequest of the object in your will. This can avoid confusion, bad feelings, or dissention among children when they divvy up your possessions. Also, do not be offended if something particularly special to you is not requested by one of your children. Try to be as objective as possible.

The estate-planning team should include your consulting forester, an estate-planning attorney, an insurance underwriter, your personal representative or executor, and, possibly, one of your children. An accountant may also be part of the team, if necessary. Estate planning is a relatively new and somewhat specialized type of law practice. Chances are good that the local attorney who has always handled your work will not have the expertise to draw up and execute the necessary documents. If the attorney is also a friend, there may be bad feelings when you hire someone else for the estate plan. Under no circumstances should you feel obligated to provide legal work to anyone for reasons other than competency. In this area, even small errors are costly, and you will not be around to help clear things up. Ascertain that the attorney you work with has extensive experience in estate planning, preferably where woodland assets are involved. A good way to obtain a reference to an attorney is to inquire at your state or local land trust. Attorneys who have worked with land trusts usually have had experience with estate planning.

A good source to help locate a qualified estate planner is The American College of Trust and Estate Counsel (3415 S. Sepulveda Boulevard, Suite 330, Los Angeles, CA 90034, 310-398-1888). The college requires members to have a minimum ten years' experience in estate planning and sound local and national references. Not every estate planner is a member of the college.

Your *executor* is the person named in your will whom you trust to handle business decisions, pay bills, and disperse your belongings to the people you have designated in your will. Anyone can serve as executor, even your attorney. Often, surviving spouses or children are appointed to represent the deceased. There are many points to consider in picking an executor that are beyond the scope of this book, but you definitely want someone who understands your commitment to the forest and is willing to ensure your requests are fulfilled—at least during their time of appointment. If you decide to avoid probate through the formation of a living trust (discussed later), you need to appoint a trustee, who can also serve as your personal representative.

If keeping your forest lands healthy, productive, and intact is a priority, you need to locate a consulting forester who understands your goals and is capable of achieving them. This may or may not be the person with whom you have worked in the past. The consultant must be someone you believe in and trust as much as you trust your personal representative or trustee. If you have any misgivings about consultants you have worked with in the past, it is time to shop for a new advisor. If your heirs are not apt to be up to speed about making management decisions after you have died, develop an agency relationship with the consultant, and create the necessary documentation that allows the consultant to act for your trust or your estate when you are gone. Because it is legally impossible to act as an agent of a recently dead person, it is essential the consulting forester is identified as an agent to your estate, much as you will appoint someone to be your personal representative or executor to handle your personal and financial affairs and to disperse your estate.

One of the first things the consultant will do is update the management plan, complete an appraisal of the woodlands, and document your ideas for the future forest. The woodland appraisal will probably need to be accepted and certified by a state-licensed appraiser. This is usually not a problem if the numbers in the appraisal seem reasonable.

Perhaps one of the most difficult decisions will be designating a family member through whom the consulting forester communicates with others who have an interest in the forest—at least until your affairs are settled and the estate has been dispersed. You may opt to have the consultant communicate through the executor. Or you may want to appoint one of your children or other trusted heir to work directly with the consultant to implement your plans; He or she may well also be the person who will inherit the land (less any rights you might have sold, donated, or bequeathed for conservation purposes).

### *Lowering Estate Value*

Estate planning to most people means avoiding estate taxation. If the value of your assets exceed the limits discussed in chapter 8, your estate will owe an estate tax. An important part of the estate-planning process is figuring out the fair market value of your estate, or approximately what you think it might be worth on the date you (or your surviving spouse) will pass away. Obviously, it is impossible to be precise, and no one expects you to die as "planned." Make a good guess, though, because if the value exceeds the limits already discussed, your heirs will pay estate tax.

If you know the total estate value exceeds the tax-free limits allowed in the Unified Gift and Estate Tax Rules, you can begin lowering your estate's value by making tax-exempt gifts, charitable contributions, giving or willing certain rights to conservation organizations, and by other means. You and your spouse can also establish the special tax-saving forms of a living trust to avoid taxation on the portion of the surviving spouse's estate that exceeds the maximum exemption. Remember, the surviving spouse can inherit an estate of unlimited size; but when that spouse dies, only one exemption is available. If the exemption available to the first spouse to die is not used to set up a trust, it is permanently lost to the family's estate. Amounts that exceed the exemptions are taxed at rates beginning at 37 percent.

A living trust can be designated to divide the family's assets into two trusts, thereby taking advantage of the exemptions available to husband and wife. It is an excellent way to beat estate tax (or at least lower the tax), avoid probate, and keep forest land in the family. (Trusts are discussed in more detail later.)

*Sharing Property with Children.* When title to property is held between two or more people as "joint tenants with rights of survivorship" (often abbreviated on documents as "JTROS"), the surviving joint owners automatically own 100 percent of the property if one owner dies. An interest that someone has in a JTROS property title is not part of the person's probate estate. Whatever interest the decedent had in the property automatically reverts to the survivors. Consequently, many families mistakenly use this strategy to remove assets from a parent's estate to avoid estate taxes on the property. Unless the children can prove they contributed to the formation of the joint tenancy, or the joint tenancy was formed as a result of a long-term, tax-exempt gifting strategy, the full value of the JTROS title is included in the decedent's estate. In other words, simply changing the property title to include children is not a valid strategy to lower estate value. Also, when you share title as joint tenants, you give up control of the property, and a joint tenant cannot sell his interests without consent of the others (state laws vary in this matter). Fortunately, a joint tenant's share is usually not very marketable.

Another, more acceptable way to share title with children and lower estate values at the same time is discussed in the section on family partnerships.

*Personal Gifts.* The IRS allows a single taxpayer (donor) to make a tax-exempt, annual gift of up to $10,000 per individual (donee) to an unlimited number of individuals. A married couple (considered to be two taxpayers by the IRS) can give $20,000 in tax-exempt gifts per donee. So long as the gift is less than these amounts, no federal gift tax is due (although state gift tax may be due). Also, the gifts do not reduce or consume the Federal Unified Gift and Estate Tax Credit otherwise available. The gifts are entirely exempt from federal taxation. A gift can be cash or anything of value, but it must be given unconditionally and be immediately negotiable for its fair market value on the date of the gift. In other words, the gift must be worth the amount claimed when the gift was made, and the donee can do anything he or she wants with it. Also, the Taxpayer Relief Act of 1997 will increase the annual tax-exempt gift amount according to an inflation index.

Conceivably, a married couple owning forest land could give undivided interests in woodlands to their children in the amount of $20,000 per child. If those children are married, the gift can be as high as $40,000 to each married couple (that is, $20,000 per donor spouse to each donee son and his wife, who is also a donee). However, since the gift must be unconditional, this may not be the best way to handle forest assets. A cash-strapped son might be forced to sell his gift despite your wishes. There is also the possibility of divorce, and remember that the gift carries with it whatever basis you have in the property (see chapter 8). If your cost basis in forest land is small in proportion to current fair market value, your family may be better off to inherit forest land rather than receive it as annual gifts.

There is a way to make annual tax-exempt gifts of forest land to children while maintaining control over the assets (discussed in more detail below in the section on family partnerships).

*Charitable Contributions.* Another way to lower estate value is to make charitable contributions to qualified organizations. There are three types of qualified organizations: (1) public charities, such as universities; (2) semipublic charities, which are not controlled by state charter; and (3) private charities.

Most charitable organizations, public and private, can accept gifts and ensure tax benefits to the donor. Private charities must meet the requirements of IRS Code Section 501(c)(3)—basically a not-for-profit, charitable, educational organization—to offer tax-savings benefits to donors. The fair market value of gifts to qualified public charities can offset up to 50 percent of the donor's adjusted gross income (AGI) in any single tax year. The value of gifts that exceed 50 percent of AGI can be carried over into succeeding tax years for up to five additional years. The deduction can be limited to 20 or 30 percent of AGI per year for certain gifts, but the carryover rules are the same.

Gifts can take the form of cash, art, jewelry, stocks and bonds—anything of value. Although the strategy for a forest owner might be to lower estate value by giving nonforest assets to keep the woodland resources intact, it is not uncommon to give an easement "for conservation purposes." The "benefactor" of the donation is usually a land trust. Any qualified conservation organization, however, can accept the donation of an easement. An easement given to a neighbor does not qualify as a charitable contribution. (The details of qualified charitable contributions are discussed later in this chapter.)

Some of the advantages of giving an easement to a land trust are the following:

1. Estate values are dramatically lowered because the easement eliminates the proportion of fair market value of land that is attributable to its potential value for development.
2. Because the easement is usually a transfer of development rights, the land is protected from development but is still owned by the family and can be sold or passed to new owners (with the easement still intact).
3. The donor can use the value of the gift to offset income for up to six years, thus lowering income tax liability.

The tax advantages of giving an easement are thus twofold: it lowers income tax liability while you are alive, and it can lower or eliminate estate tax liability after you are gone. Because the fair market value of the property has been lowered by the value of the easement, property taxes should be lower. This is not often the case, though, and the point has not been argued enough in courts to have established a precedence. Still, if you have given an easement that encompasses development rights, your property assessment should be substantially lower. Taxing authorities are reluctant to lower the assessment on protected lands because land trusts do not pay taxes on easements they hold. The nature of the easements means they have no market value, since the trust cannot sell the easement. From the town's perspective, it is as though a portion of its grand list has evaporated. Most authorities on the subject agree, the dilemma of how to tax protected lands will be resolved as more and more communities address the question of fair taxation on farm and forest lands.

*Deferred Gift.* A gift to a qualified charitable organization can also be deferred. The deferment can be specified many different ways. For instance, the gift can be effected immediately but with a provision that you (and your survivors, if you wish) receive income from the property even though the title of the property has passed. A gift of a remainder interest means you continue to own and enjoy the benefits of the property while you are alive. When you die, whatever is left of the property (i.e., the remainder) goes to a charitable organization. A testamentary gift is any gift effective on death. When made to a qualified charitable organization, the value of the gift is fully deductible when figuring the total taxable estate. The testamentary gift has estate tax advantages but does not have current income tax savings (table 9.1). This is because the donor can change his or her mind (rewrite the will) before death. A gift of a remainder interest does have income tax advantages as well as estate tax savings. Figuring the income tax advantages of a remainder interest, however, is very complicated and requires the services of a qualified estate planner or accountant.

A variation on the gift of a remainder interest is a present gift of a future interest in property. This is an especially useful technique when the gift involves real property, such as woodlands. The charitable organization that accepts the gift usually wants to have a large degree of control as to how the property is used during the lifetime of the donor (or the other beneficiaries if the future interest extends beyond the lifetime of the donor). In exchange for control, the charitable organization—usually a land trust—will hold the property in trust and pay the donor an annual fixed or variable annuity that is based on a percentage of the fair market value of the assets. When the donor or designated beneficiaries pass away, the

*Table 9.1. Tax Implications of Different Strategies to Lower Estate Value*

| Option | Income Tax | Estate Tax |
|---|---|---|
| Easement gift | Savings | Savings |
| Remainder interest gift | Savings | Savings |
| Testamentary gift | No savings | Savings |
| Give to children | No savings | Savings |
| Will to children | No savings | No savings |
| Testamentary gift of retirement assets (IRAs, 401ks, etc.) | Savings | Savings |

*Source*: Adapted mainly from Small 1992.

property belongs to the charitable organization. These types of gifts must be irrevocable, so they require careful thought and planning. Gifts of a remainder interest are often used when the donor does not have any direct descendants and he or she is concerned about the cost of elder care and/or a protracted illness. Organizations that accept present gifts of future interests are usually very flexible about the terms of the annuity. They are more than willing to design support arrangements that ensure the donor that costs will be met during his or her lifetime, and will usually also accept the donor's conditions about how the land is to be managed and used by future owners.

*Special Valuation.* Internal Revenue Service rules allow farm and forest lands to be evaluated for estate tax purposes using special valuation procedures. These procedures allow lands to be assessed at current-use values rather than fair market value. Often, these rules are discovered by the family after the deceased has passed away, and the federal estate tax is due within nine months. If it is your family's intention to keep land intact and to continue managing the woodlands, you may want to investigate special valuation. Recapture rules apply, however, if the land is used for purposes other than what was specified in the special valuation. The services of a qualified estate planner and an independent appraiser are necessary to claim special valuation of forest resources.

The Taxpayer Relief Act of 1997 also allows the executor to exclude up to 40 percent of the fair market value of forest and farmland on which an easement has been given to a qualified organization. (For details, see the section on Estate Taxation in chapter 8.)

## Family Partnerships

A partnership is a noncorporate association of two or more people, each of whom owns a share of an undivided interest in property. An undivided interest is similar to the shares of a corporation traded on the New York Stock Exchange. When you buy stock, you are buying shares of the corporate assets, not any one part of the company. The concept of undivided interests as it relates to family partnerships involving forest land is perfect because it does not require a specific designation of woodlands to any one of the partners: It avoids the need to split up the land among the children.

A family partnership can be a form of limited partnership where there are two types of partners: (1) general partners who make all decisions—you and your spouse—and are responsible for day-to-day affairs and (2) limited partners—your children.

Although limited partners own an undivided interest in the assets, they have no authority to make decisions. Also, the children are usually asked to sign agreements that allow the other partners to buy out an individual's interests should he or she decide to leave the partnership.

How do children obtain an undivided interest in the forest? Through annual, tax-exempt gifts from the parents (described earlier in this chapter). Each child can receive up to $20,000 in annual tax-exempt gifts from a husband and wife. And the best part is this: The actual gift is an undivided interest in forest land. Because a child has severe limitations imposed by the partnership as to the marketability of his or her interests, the parents can transfer $30,000 of value to create a $20,000 gift. This is known as discounting, and it is one of the primary reasons family partnerships are so attractive as a forest estate-planning tool. Discount rates of between 30 and 35 percent are considered reasonable, though rates as high as 70 percent have been argued in tax court.

It is significant to note that the Internal Revenue Service gives careful scrutiny to family partnerships designed solely (and obviously) to avoid estate taxes. Excessively high discount rates are the flag. Special care and good, sound legal advice is essential.

A family partnership allows parents to disperse the estate but keep the forest assets intact. Because the estate has been dispersed to the partnership, little or no estate tax is due. And the parents maintain control—even if their share of the partnership is small compared to the children's shares—until new general partners are appointed. The new general partner(s) is the child or children who have the greatest interest (intellectual, not financial) in the forest and the best ability to carry on in your traditions. The other children share income and other benefits of owning forests, but they make no decisions.

A new variation on the theme of the family partnership is the "limited liability company" (LLC). It offers the corporate benefits of protection from personal liability and all the benefits of a family partnership. This type of association is fairly new, although it is purportedly recognized in all fifty states. For more information on LLCs, talk with an estate-planning attorney.

*A Gifting Strategy Using a Family Partnership.* Consider the case of a husband and wife who own forest land that is worth $950,000. They have four children and approximately $450,000 in other assets—stocks and bonds, primary residence, life insurance, autos, antiques, jewelry, and other personal belongings. Their total estate is $1.4 million. Even if they create living trusts designed to minimize estate taxes (described later), they can shelter only $1.25 million from estate tax in 1998, and up to $2 million in 2006. To be on the safe side—if one of them dies unexpectedly—they decide to immediately lower the estate value by at least $200,000. Because their assets are apt to appreciate over the years they expect to live, however, they wisely plan to lower the estate by $500,000. They do this by forming a family woodlands partnership, designating themselves as general partners.

The children are brought into the partnership as limited partners. Since the IRS has consistently accepted a 30 percent discount rate on assets transferred from parents to children—who must accept the assets according to the terms of the partnership agreement—the parents can annually transfer, say, $28,000 to each child for a tax-exempt gift of $19,600 ($28,000 less 30 percent is $19,600). With tax-exempt gifts to each of four children, the parents can lower their estate value by $112,000 each year. In five years they will have reached their goal. Assuming the balance of their estate is in living trusts, no estate tax is due when the surviving spouse dies.

Before the surviving spouse passes away, a new general partner is appointed. When the surviving spouse passes away, the partnership continues. Children who want to leave the partnership are bought out by the other partners, according to the terms of the agreement. Your goals of maintaining a long-term management strategy and keeping the land in the family succeed, and your children (and their children) applaud your genius.

## *Living Trusts*

A *trust* is an arrangement that divides legal and beneficial interest in property among two or more people (Prestopino 1989). A trust has the following four parts:

1. The *grantor*, also known as the settlor, donor, creator, or trustor. This is the person with the property who initiates the trust and establishes the directions governing the administration of all trust property.
2. The *corpus*, or body of the trust, also known as the principal. This is the property the grantor has transferred to the trust.
3. The *trustee*, which is a person or other entity agreeing to hold legal title to the property for the benefit of others.
4. The *beneficiaries*, or the people who benefit from the trust without having any legal ownership.

Trusts are usually formed as a means of providing for individuals who may not be capable of managing the property themselves. It is also a way to conserve property so it is not squandered by beneficiaries who might otherwise be compelled to do so.

A *living trust* is created during your lifetime. Like a will, it includes directions for the disposition of assets in the trust both during your lifetime and upon your death (Bove 1991). In a living trust, the grantor is also the trustee and the beneficiary. In other words, income from property held in the living trust is paid to the grantor, and all the property remains in the grantor's complete control. When the grantor passes away, trust income is paid to survivors, or the property in the trust is dispersed to heirs. Because the trust involves a legal transfer of a person's property to the trust, and the trust spells out what is to happen to those assets, there is no need for probate. When the grantor is also the trustee, a co-trustee and/or successor trustee is usually appointed. That way when the grantor dies, the trust is not left without a trustee.

So what is the benefit of a living trust to a forest owner? A living trust can be designated to double the amount that can pass free of estate tax to the children of a married couple. This type of trust is referred to as an "A/B trust," a "credit-shelter trust," or a "family trust."

Assume the "owner" of such a trust is actually a married couple, and the 1998 value of their forest land and all their other assets is $950,000. The wife dies in 2000. Assume that in two years their 1998 estate has appreciated at an annual rate (compounded) of 7 percent, so their estate is worth about $1.1 million. Current rules allow the surviving spouse to inherit the entire estate without paying any taxes. But when the surviving spouse dies, depending on the exemption amount that year, the balance (technically, from the first spouse's share and any amounts that exceed the exemption available to the recently demised spouse) is taxed.

Remember, the estate assets are appreciating at 7 percent, while the exemption amount (under the 1997 Tax Law) is increasing by an average of about 4 percent. If the surviving spouse dies in 2006 (six more years of compounding the total estate at 7 percent), the estate is worth about $1.6 million. The Unified Gift and Estate Tax Rules would permit $1 million to pass free of estate tax, but the family will pay estate taxes on the $600,000 that is not exempt.

If this same couple had transferred their assets to two A/B trusts, one in the name of the husband and the other in the name of the wife, the family could have avoided federal estate tax altogether (table 9.2). By dividing their combined assets into two trusts, they can take full advantage of the Unified Gift and Estate Tax rules. In 1998 the couple can shelter up to $1.25 million, and in the year 2006 they can protect up to $2 million. During the lifetime of the couple, all income from property held in the trust is paid to them or to the surviving spouse. When the surviving spouse dies, the trust is dispersed to heirs, or it can continue with the children or others who are designated to benefit from the income.

The particulars of living trusts are far beyond the scope of this book. Many excellent references have been written on the subject (see the References and Suggested Readings section). A word of caution: Although it is possible to create your own living trust, it is tricky business and best left to a qualified estate-planning attorney. Properties need to be retitled to the trusts, and you must follow IRS

*Table 9.2. Formation of a Family Trust and Projection of Its Effect on Estate Taxes*

| Year | Event | Status | |
|---|---|---|---|
| 1998 | A/B trusts formed | Husband and Wife Estate $950,000 | |
| | | Husband Trust $475,000 | Wife Trust $475,000 |
| | | *(Assets appreciate at 7% compounded per annum.)* | |
| 1999 | | Husband Trust $508,000 | Wife Trust $508,000 |
| 2000 | Wife dies | Husband Trust $544,000 | Wife Trust $544,000 |
| 2001 | | *(Both trusts continue to appreciate. Income from both is available to* | |
| 2002 | | *the surviving husband.)* | |
| 2006 | Husband dies | Husband Trust $763,000 | Wife Trust $763,000 |
| 2007 | Estate tax filed | Total exemption available to both trusts: $2,000,000 Estate Tax Due: 0 | |

rules to the letter. It may cost from a few hundred to a few thousand dollars in legal fees to set up the trusts (expect to pay between $1,200 and $2,000), but the tax savings can be astronomical (depending on the circumstances, estate tax savings in excess of a few hundred thousand dollars). Also, a trust avoids probate expenses of $3,000 to $10,000.

## Land Trusts

Over the past thirty years, hundreds of mostly local, private, non-profit conservation organizations have emerged to protect important forest and farmlands from development. Land trusts, as these organizations have come to be known, grew out of a need to provide donors with assurances that lands could be protected long after they had died. The word *trust* has the same meaning here as applied elsewhere: an arrangement that divides legal and beneficial interests in property. The word *protection* usually means from development into nonforest or nonfarm uses. The purpose of land trusts is to give individuals a choice about how lands will be used when passed by sale, gift, or bequest. The land trust, with respect to your forest land, is the equivalent of the personal representative of your estate, with one important difference: A land trust is a legal entity that has been carefully designed and formed to exist in perpetuity. Even if the organization you work with goes out of business fifty or one hundred years from now, a carefully crafted chain of successors will take on the obligation of always and forever protecting the property's title and your legacy to the land.

The most common method of protecting forest land is to give an *easement* (see chapter 3) to a state or local land trust. Also known as a *transfer of development rights*, the easement is designed by the donor or grantor to specifically allow certain activities while disallowing others. Although the easement may disallow development, it can specify certain areas where family members (present and future generations) are allowed to build if they so desire. The easement can be as general or as specific as you—the grantor or donor—require, within the limits imposed by the IRS. The more specific the easement, however, the more difficult and expensive it is to enforce.

A land trust will sometimes buy an easement to protect an especially significant tract, but usually it will only accept gifts. A donor is expected to pay the costs of appraising the easement's value and the legal costs of drafting and recording the easement. Most land trusts try to recover the costs of staff involved in the transaction. Donors are usually also asked to provide an endowment, a lump sum of money that the trust invests to help cover the cost of enforcing the easement in the future. Wealthy donors are also asked to make gifts of the type described in table 9.1. For some people, the income and estate tax advantages of gifts to a land trust are very appealing. IRS rules regarding charitable contributions generally apply to the gift of an easement and any endowments.

## *Like-Kind Exchanges*

Land trusts often identify parcels that are important for connecting wildlife travel corridors, for protecting water resources or prominent vistas, or for adding to an existing block of protected land. When owners of these lands are approached, often they are unwilling or financially unable to donate an easement. For instance, even a willing owner may not have enough current income to take full advantage of the tax savings when an easement is donated, and may be expecting profits from a sale of the land at retirement. An unwilling potential donor may not want to lose the economic potential of the land for crops, timber, or future development. For both willing and unwilling potential donors, a like-kind exchange with a land trust may prove acceptable.

A like-kind exchange is a tax-free transaction, usually initiated by a land trust, whereby an owner exchanges his or her property for qualified, like-kind property. As long as the like-kind property qualifies under IRS rules, there is no taxable gain. The advantage to a landowner is the ability to defer the capital gain that would otherwise be due with an outright sale, and to obtain property of like-kind that allows fulfillment of financial goals with minimal impacts on important landscape features. The advantage to the land trust—and society—is protection of significant lands from development. For more information on like-kind exchanges, contact your local land trust.

A common misconception is that land trusts buy up land and lock it away. Nothing could be further from the truth. A land trust only rarely becomes involved in outright land purchase, and then briefly—only just until it can find another buyer who is willing to purchase the property with an easement. Land trusts mostly accept gifts of easements (which are of no value to the trust, since it cannot resell the easement). They exist on membership dues and contributions, foundation grants, special state appropriations, and outright gifts of land and other assets. When a land trust accepts an easement from a woodland owner, it expects the owner to help draft the easement and to have clear plans as to how the land is to be used. Family descendants can inherit the land, and they can carry on the original forest-management plans. They can also sell the land to a buyer who is willing to accept the title with easements. And a family can reserve areas for house lots if that is acceptable to the land trust.

Land trusts do not dictate how forest lands are to be managed. Rather, they negotiate an appropriate and acceptable management strategy with each owner and design an easement that ensures those practices are followed. It is the current owner's responsibility to follow the plan, and if he or she strays from the plan, the land trust will intervene. What constitutes straying from the plan is spelled out in the easement. Most trusts will not quibble over small changes in management. But drastic alterations to the landscape, houses going up, or any violation of the easement will cause the land trust to investigate. An owner of protected forest lands should seek guidance from the land trust that holds the easement before implementing management practices that drastically alter the landscape. Also, as part of the land trust's perpetual monitoring responsibility, a representative from the trust periodically will visit to ensure the easement is intact.

After working with a local land trust, some owners decide to leave a remainder interest in their land to the trust (as described above in the section on charitable contributions). The tax advantages—both of income tax savings and lower estate taxes—are often very attractive. This is especially true for owners without children, or with children who are not interested in owning the land. When a land trust accepts land as an outright gift, or a gift of a remainder interest, it will usually separate the development rights and then sell the land (either during the life of the donor or after, depending on the circumstances) to someone who is willing to honor and protect the easement. Any profits from the sale are used to continue its good work. Under certain circumstances the land trust will set up an annuity, the beneficiary of which is the donor until he or she dies. The trust will also make arrangements for elder care and other expenses the owners are apt to sustain during the balance of their lives. Most land trusts are highly flexible in the design of easements and financial arrangements as long as their primary concern is met: to protect land from development.

For more information on land trusts in your area, contact The Land Trust Alliance, 1319 F Street NW, Suite 501, Washington, DC 20004-1106, 202-638-4725, www.lta.org.

## Locating an Estate-Planning Attorney

The best way to locate a suitable estate-planning attorney is to make inquiries about his or her practice. You want someone who devotes at least half-time to estate planning, which may entail preparation of five or more estate plans each month. Ask if they are involved in continuing education seminars. Because estate planning is constantly changing, active involvement in professional development in this area is essential, at least to the extent of ten or more hours per year.

Find out if the attorney has given presentations to groups on the subject and, if so, can provide you with a copy of the materials used. Most estate-planning attorneys are asked to speak a few times each year. A copy of the teaching materials will give you hints as to their focus and how well organized and experienced they are.

Another key question: Does the attorney prepare his or her own standard forms for wills and trusts or obtain them from another source? How often are the forms revised or updated? Obviously, the attorney should have his or her own forms, and updates should be continuous to reflect changes in the tax law or changes in local probate procedures or statutes.

Finally, ask if the attorney has had any experience working with forest owners, especially where the disposition of forest assets was a major consideration. Have they ever worked with a forester and, if so, on what types of projects?

The attorney may be able to provide references, but is bound by rules of confidentiality from revealing the identity of a client, let alone discussing the specifics of another client's estate plan. However, you might ask if you may have permission to speak with at least one recent client.

## Summary

People who own forest land have a special responsibility that extends beyond a lifetime. It is this responsibility, more than the financial gains, that make estate planning an essential exercise for all families owning and tending woodlands.

Following are some points to remember about planning for forests in your estate:

- Do not be fooled into thinking the best way to avoid estate tax is to leave everything to your spouse. Eventually, the estate of one or the other may have to pay a tax, which starts at 37 cents on the dollar.
- Obtain advice from a qualified estate planner on when to use joint tenancy with rights of survivorship for personal property, such as automobiles and bank accounts. Use JTROS to share woodland ownership only with the advice of an estate planner.
- Know the value of forest land in your area. It may be higher than you think—high enough to trigger an estate tax your family will not be able to pay.
- There are lawyers, and then there are lawyers who know estate planning. Choose the latter.
- Involve your children in estate planning; find out who is interested in maintaining the woodlands and who can carry on your traditions.
- Learn more about every angle of estate planning but *don't do it by yourself!* Hire an experienced estate-planning attorney to draw up the necessary documents.
- Assemble a team that includes a consulting forester, an attorney, an accountant, an insurance underwriter, and interested family members.
- If your total estate exceeds estate tax credit limits, investigate ways to lower the estate value.
- Consider the advantages of living trusts as a way to hold assets to avoid probate and to minimize estate taxes.
- Think about giving an easement to a local land trust to gain immediate income tax advantages, lower estate value, and ensure woodlands are protected from development.
- Finally, consider a like-kind exchange of land with a local land trust to forever protect an important feature of your forest.

The forest owner's estate plan should emphasize descriptive phrases and ideas you believe are important about forest values that should be recognized, managed for, protected, or celebrated in the future. It is your chance to create a living legacy by which people will remember you long after you have passed away. The plan should acknowledge uncertainty and be flexible. It should describe your visions, the principles behind decisions you have made in the past, and the conditions you believe are desirable for the future. Finally, the forest estate plan is your chance to leave behind something truly important.

## Further Reading on Estate Planning

For more information on estate planning, consult these publications, listed in References and Suggested Readings at the back of this book: A. A. Bove, Jr., *The Complete Book of Wills and Estates*; P. C. Kaufman and S. H. Green, *Understanding Estate Planning and Wills*; H. L. Haney and W. C. Siegel, *Estate Planning for Forest Landowners*; C. J. Prestopino, *Introduction to Estate Planning*; S. J. Small, *Preserving Family Lands: Essential Tax Strategies for the Landowner;* M. Dowd, *Estate Planning Made Simple;* and D. T. Phillips and B. S. Wolfkiel, *Estate Planning Made Easy.*

# Chapter 10

# Settling Disputes and Shopping for an Attorney

Occasionally, a forest owner is faced with a dispute of one sort or another. Unless you are an especially litigious person, your goal should be to avoid disputes. When a dispute is unavoidable, try to settle as reasonably, quickly, and amicably as possible. This may sound like a simplistic approach to what could prove to be an otherwise complex and vexing problem you share with difficult people, but using an attorney and court proceedings to solve a dispute should be viewed as a last resort. Aside from the expense and uncertainty of asking a court to help settle a dispute, most people are highly discomforted by this type of encounter with the law. Anxiety, sleepless nights, distraction, and unease are common complaints of people who find themselves party to a lawsuit. Former litigants often describe a protracted court battle as the worst experience of their lives. Even if you are certain your claims will prevail in court, the emotional expense of the battle may not be worth vindication. Sometimes it is easier and cheaper to give in than to fight.

The purpose of this chapter is to describe common circumstances around which woodland-related disputes occur, and to discuss various means of resolving disputes—from attempting to resolve the problem yourself to hiring an attorney and going to court. The emphasis is on resolving disputes without a fight, and avoiding court.

Whenever two or more people cannot resolve their differences and a resolution of some sort is important, it is a dispute. In other words, people can disagree on something that is a matter of opinion and walk away with the issue unresolved—they can "agree to disagree." The outcome is not a matter of consequence. The word *dispute* is used here in the context of a disagreement where the resolution is important.

## Common Areas of Dispute Involving Forests

Some areas where disputes involving forest lands are common (and important) are as follows:

*Boundaries, Surveys, Rights-of-Way, and Deeds.* Disputes involving these areas almost always require the services of a licensed surveyor and an attorney. Any dispute regarding a deed should be handled by an attorney. Unless you know what you are doing, you should not attempt to resolve questions about a deed by yourself.

A boundary dispute may be easily resolved with respect to the concerns of the current titleholders, but when one of the owners attempts to pass title, the dispute requires formal resolution. A dispute may arise as to who will pay for professional services to help resolve the question. State laws may prescribe remedies for these types of disputes, so it is probably well worth the money to get a legal opinion.

An owner may grant a temporary right-of-way to allow a neighbor easier access to his or her lands. In some states, an owner is required to grant a temporary right-of-way specifically for forestry purposes. Sometimes the location, duration, and other relevant aspects of the right-of-way cause a dispute. These can usually be resolved by the owners without further assistance. However, if you are asked to give a right-of-way and you are reluctant to do so—for any reason—it is best to seek legal advice.

*Trespass.* Trespass is a dispute with someone who is using (or has used) your land without your prior approval. It can usually be handled without the assistance of an attorney. Be sure your land is properly posted (see chapter 3), and—if necessary—follow up a personal conversation with a letter to the trespasser. As a concession, consider allowing access but limiting the scope or period of access.

If the trespass has resulted in damage or loss, unless you know the laws in your state, it is best to seek legal advice. In some states, damages are automatically awarded a woodland owner for timber trespass; for example, a logger knowingly crosses your boundary and cuts timber. Getting the errant logger to pay up is another matter. In cases of trespass, or suspected trespass, speak with the trespasser—get the story—before having threatening letters sent from your attorney.

*Liability and Personal Injury.* These types of claims should be handled by an attorney. When people perform work for you, they should provide evidence of general liability and workmen's compensation insurance (chapter 6).

*Timber Sale Terms and Professional Service Contracts.* Disputes involving these types of contracts often end up in court, needlessly. As discussed in chapter 6, it easy to build provisions into these types of agreements that allow for alternative dispute resolution methods, such as arbitration or mediation, described later. The best way to avoid these types of disputes is clear, written communication. Also, do not view the contract as a means of forcing someone to do something not provided in the contract. That simply sets the stage for a dispute. If you don't trust the person, don't sign the contract. Find someone else to work with.

*Divorce.* When forest land is one of the assets in a divorce settlement, too often the land suffers a quick sale at bargain prices to a buyer who cares little for sustaining long-term forest benefits. Although resolving the larger disputes that spawn divorce are far beyond the scope of this book, spouses who care for the land can at least agree to protect it when they split up. For instance, one spouse can agree to buy the other out over time; or they may agree to give the development rights to a local land trust (and split the tax benefits). Or they may want to create a trust for their children (see chapters 8 and 9).

Generally, it is better to resolve issues about the forest *before* one partner sues the other for divorce. A happily married couple can execute an agreement about the disposition of woodlands "in the event of a divorce." Although this may sound like anticipating divorce, at least the partners agree: The importance of the forest transcends any future disputes that might come between them as a couple.

Resolving the issue of forest assets in the event of a divorce requires clear, written communication between spouses, and—since the resolutions proposed here are somewhat unusual—may require the services of an attorney to draw up the agreement.

When one prospective spouse is bringing forest land to the marriage, the parties should have a prenuptial agreement. This is especially important in the common-law states (which is every state except the following community-property states: Arizona, California, Idaho, Louisiana, Nevada, New Mexico, Texas, Washington, and Wisconsin). In a common-law state, a family court judge decides who gets what based on a wide range of factors, such as length of marriage, earning capacity of the individuals, fault, and other issues. Without a prenuptial agreement, all assets of the couple—even those brought to the marriage—may be considered joint assets.

In a community-property state, the assets a spouse brings to a marriage are usually not part of the marital property. However, assets acquired during the marriage are divided fifty-fifty, unless there is a prior agreement that specifies some other distribution. An agreement drawn up after the marriage is called a *postnuptial agreement*.

*Taxes.* A dispute over taxes usually pits you against the system. In the case of property taxes, as noted in chapter 8, you must demonstrate a mistake in your assessment or an error in calculating the tax. Forget about arguing the larger issue of fair taxation. If you do have a valid dispute, you may be able to argue it successfully without professional assistance.

Sometimes a well-informed taxpayer can handle a state or federal audit regarding income and expenses from forest lands without the services of an attorney or an accountant. If the audit involves a much broader examination—if the income from a timber sale is incidental and not the main issue—you may want to at least have your figures checked by an accountant for the year in question before going into the audit.

## Solving Disputes on Your Own

As a forest owner, there will be more instances where you should attempt to resolve a dispute on your own than instances that will require professional services. The trick is knowing when to ask for help. If you are the type of person who avoids conflict at all costs, you may want to ask someone to represent you. If you are a person who enjoys arguing and never gives up a good fight, you too should ask someone to represent you. The biggest mistake an individual makes in confronting others is letting his or her ego get in the way. Never, under any circumstances, should a valid dispute between individuals be taken to a personal level. That only causes the parties to lose sight of the issue at hand, which gets lost in the fray. Animosity builds rapidly, all reason is lost, and there is no hope of reaching an easy solution. In fact, the objective often shifts from resolving the dispute to destroying the other person. It is impossible to negotiate with someone you detest.

If you know yourself well enough to admit you might be closer to one extreme or the other, you can best serve your cause by hiring a consulting forester, or some other appropriate professional advisor, to serve as your personal representative. Be wary, though, of legal advice from someone who is not qualified to give it.

Assuming you are of a mind to negotiate your own resolution to a dispute, the first objective is to identify the problem three different ways: (1) from your own perspective; (2) from the perspective of your adversary; and (3) from the perspective of a third party who knows only the facts (and not the personalities of the adversaries).

This exercise is not as simple as it sounds, but you may discover that it is easier to understand the problem and resolve it than you first imagined.

Decide what you are willing to concede to resolve the dispute. This may include something you are willing to give even though you do not believe you should be required to do so. Remember, the alternative is an escalating battle, expensive attorneys, sleepless nights, lost work time, ill will, and the chance you might lose.

When you meet with your adversary, state your position on the issue(s) clearly and concisely. Listen to the opposing viewpoint, and allow ample time for the other side to state their points. Ignore any careless references that may begin to take things to a personal level. Now, restate the opposing position as you understand it. Give your adversary a chance to clarify, if necessary, and to restate the position again. Keep doing this until you can make a statement about their position that you understand and with which they agree. Then offer a proposal to solve the problem. Or, if the dispute does not lend itself to an immediate solution, agree that both of you at least understand each other's positions.

Follow up the meeting with a letter summarizing the situation from both perspectives. Offer a proposal to solve the problem and try to set another date to meet. You may need to negotiate among proposals and counter-proposals, and it may begin to appear too complicated to solve on your own. Do not despair—this is when you turn to alternative dispute resolution methods, described below.

If you do obtain an agreement, draft a letter describing the problem from both perspectives and the resolution of the items on which you have agreed. Send two copies, and have the other person sign one and return it to you. Signatures may need to be witnessed and/or notarized on this type of agreement in your state. Ask your attorney, or call your Attorney General's office (Appendix B). If the settlement involves more than a few thousand dollars, you may want to have an attorney review the agreement before signing it.

## Alternative Dispute Resolution

A term coined by trial attorneys, *alternative dispute resolution*, describes a process that attempts to keep disputes out of court. Though attorneys usually represent the parties who agree to this form of resolution, the parties can do it on their own. The result is usually a faster, more comfortable, and cheaper alternative to using the courts. Alternative dispute resolution methods are relatively new, and states view the proceedings differently. If you have a question as to whether a dispute can be satisfactorily resolved out of court, check with your lawyer or the Attorney General.

### *Mediation*

The purpose of mediation is to bring the parties of a dispute together and, through the assistance of a third party—the mediator—develop the terms of an agreement that resolve the problem. It is the mediator's job to focus on the problem and its resolution, to keep tempers cool, and to negotiate a solution that is mutually acceptable. However, the parties are usually not bound to the results of mediation unless the product is a properly executed and signed agreement.

The mediator can be anyone, but preferably it is someone with proper training and previous experience. It should be someone with an excellent capacity to listen, to interpret and articulate the exact nature of the dispute, and to help the parties craft their own resolution to the problem. A good mediator is nonjudgmental; it is not the mediator's position to say what is right and what is wrong. A successful mediation is when the parties to the dispute do all the work, and the mediator just keeps the discussion on course.

### *Arbitration*

In arbitration, the parties to a dispute usually agree to accept—before the process begins—the findings of an arbitrator, or an arbitration panel. It is akin to a private court proceedings, and may be handled in much the same way. One obvious advantage of arbitration is this: Unlike court proceedings, which are a matter of public record, the arbitration record is private.

Although it is not necessary to have an attorney or other representation in arbitration, depending on the nature of the dispute (and the rules regarding arbitration proceedings in your state), you may want to have an attorney present. Anyone can serve as an arbitrator, and it is not uncommon to include a clause in a timber sale contract to resolve disputes using an arbitration panel (see chapter 6).

The results of arbitration are usually binding on the parties, who are required to sign an arbitration agreement before the process begins. A disadvantage of arbitration is there are usually very few grounds for appealing the arbitrator's decision. Although state laws in this matter vary, a court judge will usually accept the decision of an arbitrator as though it were his or her own decision. It is not surprising that many people who offer arbitration services are former judges. Arbitration is an excellent way to settle a dispute. But if you believe you cannot live with a decision that goes against you, you probably need to hire an attorney and prepare for court.

## Shopping for an Attorney

Few people make it through life without at some time requiring the services of an attorney. Fortunately, most encounters with attorneys are benign, such as when closing on a real estate purchase. For relatively simple, procedural tasks like that, almost any local attorney will do. But when a dispute is brewing, the person who handled your closing may not be the best person to hire as your advocate. Some lawyers specialize in litigation, and for a problem that is apt to end up in court, you want someone who knows the process. The same advice applies to any type of specialized legal need. Beyond title searches, there are many circumstances where any available lawyer simply will not due.

Also, if you are going to court (other than Small Claims Court, where a judge is asked to give a verdict on only monetary settlements of a few thousand dollars or less, dollar limits varying by state), forget about trying to represent yourself. Consider the words of Abraham Lincoln: "The attorney who represents himself has a fool for a client." Unless you are thoroughly familiar with courtroom procedures and the rules of evidence in your state, it is money well spent to hire an attorney if there is even a slight chance you will end up in court. Without proper representation, it is all-too-easy to lose a foolproof case on a technicality. This is especially true if you are the defendant, the one against whom claims are being made. There is no sympathy for the litigant who represents himself. A judge applies the same standards whether you know the rules or not.

The best time to shop for an attorney is before trouble starts. Not that you need to go through life anticipating court battles at every turn. Your sense of urgency, though, will get in the way of a careful, well-informed selection process if you begin shopping *after* you have a problem. It is a good idea to become familiar with attorneys in your area who have experience working with woodland issues. Ask around, list the names of a few prospective candidates. Most attorneys do not charge for an initial consultation, which is a good opportunity to learn if they have any experience with cases involving woodlands. If you do not have a specific legal problem pending, a lawyer may be reluctant to take time to meet you. Nevertheless, you can learn a great deal about someone during the course of even a brief phone conversation. You just need to know the right questions to ask.

Shop for an attorney the same way you would a consulting forester (see chapter 5). Competency should be at the top of your list, and you also want someone with whom you can communicate easily. If you end up in a dispute, the attorney will become your confidant and your advocate. He or she will speak for you in all negotiations with the other party and in court. You need someone who is easy to talk to, and who can quickly assimilate the legal aspects of your case.

Assuming you have called on your attorney to represent you in a lawsuit, you should reveal everything—including information that might be detrimental to your case. You must be absolutely truthful regarding the facts of the case. If you disclose information that you later change to "improve" (or alter) your story, your lawyer may be forced to resign. As an officer of the court, a lawyer cannot knowingly allow lies to be presented as fact, especially when stated under oath. Nor can your lawyer protect you from a lie.

Lying under oath is known as perjury and is a serious offense, punishable at the discretion of the judge. You are better off telling the truth, sticking to it, and letting your attorney work the angles in court that will turn circumstances in your favor without having to deceive. Also, remember that when you go to court, your life becomes an open and very public book. Your adversary can request permission to search anywhere a judge deems reasonable for the purposes of gathering information pertinent to the case—files from the hard drive on your computer, phone records, contents of safety deposit boxes—anywhere supporting information is apt to be found. Often the discovery period of a lawsuit is used to intimidate a litigant into settling out of court. Judges expect attorneys to try to settle suits out of court. If you cannot stand the heat, negotiating a settlement is an acceptable alternative to having your life explored in a public forum.

The procedural aspects of court are too vast to describe here. Suffice to say, it can take months or years in some jurisdictions to achieve a court verdict. All the while your lawyer's clock is ticking, and every month you receive a bill that at first blush seems ridiculously too high to be true. Lawyers are expensive. This fact alone is one of the main reasons to avoid court if at all possible.

Attorneys charge for their time in the following three ways:

1. *Flat fee*, similar to the "contract for services" discussed in chapter 6. This type of fee may be used for rote tasks, like title searching or drawing up a will.

2. *Fee on contingency.* When a client is suing another and a monetary award is expected, the lawyer will usually accept a fee contingent on the amount of the award. Thirty percent is common, but the amount is negotiable. Expect to pay office expenses and agreed-upon fees that are independent of the award.

3. *Hourly fee.* In the absence of a contingency-fee agreement, most attorneys charge for their time on an hourly basis. Billing is monthly and itemizes office expenses pertinent to your file, including the number of hours (and fractional hours) the lawyer has devoted to your case. If you call your attorney on the phone and engage in a five-minute call, you probably will be charged for fifteen or twenty minutes of time (i.e., one-quarter or one-third of an hour). At $120 per hour, a five-minute call to check on the progress of your case could cost $30 to $36. Call only when absolutely necessary, and be sure you understand the billing arrangements from the beginning.

As in forestry, there are no standard contracts. For instance, you might get your attorney to agree to a cap on fees for services, costs that he or she should be able to reasonably estimate. Setting up a living trust, preparing a timber sale agreement, or negotiating a settlement are situations where it might be beneficial to set a cap. Lawyers like to be paid on a regular basis. They do not like to carry large outstanding balances on clients. Aside from the obvious reasons for this, clients have been known to not pay up when decisions go against them.

When you receive the final bill, be aware that sometimes grounds exist for offering a payment that is less than the total. Let's say you owe your attorney $6,500, a much higher balance than most law offices will allow. Perhaps you can recall incidents during your meetings when the discussion strayed from your problems to something else, or when your attorney felt compelled to carry on at length about a related experience—any other reason to justify paying less than the total bill. You might offer $3,000. The attorney will consider your offer, maybe take it to the partners, then counter with a final bill of $4,750. She/He split the difference, and you saved $1,850. Your attorney accepts the lower payment on the strength of your arguments and on the reasonable assumption that it may take an inordinate effort to get you to pay the full bill. The settlement is the business equivalent of "a bird in the hand. . . ." However, if you owe $6,500, pay it. Do not use the methods described above to cheat your lawyer out of fees owed. You might have a difficult time finding an attorney the next time you need one.

Following are some legitimate ways to save money when working with an attorney:

- Ask if there are documents *you* can prepare to help save costs. Offer to do legwork, delivering documents to courts and elsewhere. Do not ask the attorney to perform rote tasks that you could handle.
- Always have an agenda prepared when you visit the attorney. Never allow the discussion to veer too far from it. Have your questions prepared in advance.
- Take notes to avoid asking the same question at a subsequent meeting. Consider recording the meeting on tape, but ask permission first (as your attorney must seek permission of you to record a meeting).
- Follow your attorney's advice. Others (friends and family) will offer their opinions, possibly on similar cases. But your case is unique and you are paying for advice, so you should use it. Once

a case is under way, it is unethical for your opponent's attorney to contact you. In matters related to the case, all information flows through your attorney.

- Your lawyer has an obligation to try to settle your case out of court. He or she is also obligated to communicate any reasonable offers from the opposition to you, and to make only offers that you have authorized. Clearly state your bottom line and prepare to go below it if necessary.

Finally, regard a good lawyer on your side as an insurance policy—something you hope never to use, but if you need it someday, it's there. You can do many things to avoid disputes or resolve them before they escalate into courtroom battles. But when all else fails, there is no better advocate than an attorney who understands your circumstances and your commitment to the forest.

# Appendix A. Extension Forestry Specialists in the United States

| | | |
|---|---|---|
| Alabama | 334-844-1038 | brinker@forestry.auburn.edu |
| Alaska | 907-474-6356 | ffraw@uaf.edu |
| Arizona | 520-621-7255 | cppr@ag.arizona.edu |
| Arkansas | 870-777-9702 | froth@uaexsun.uaex.edu |
| California | Declined to be listed | ——— |
| Colorado | 970-491-7780 | craigs@cnr.colostate.edu |
| Connecticut | 860-774-9600 | sbroderi@canr1.cag.uconn.edu |
| Delaware | 302-739-4811 | austin@smtp.dda.state.de.us |
| Florida | 352-846-0883 | mgj@gnv.ifas.ufl.edu |
| Georgia | 706-542-7602 | cdanger@uga.cc.uga.edu |
| Hawaii | 808-956-8708 | elswaify@hawaii.edu |
| Idaho | 208-885-6356 | mahoney@uidaho.edu |
| Illinois | 217-333-2778 | m-bolin@uiuc.edu |
| Indiana | 765-494-3580 | billh@fnr.purdue.edu |
| Iowa | 515-294-1168 | phw@iastate.edu |
| Kansas | 785-532-1444 | cbarden@oz.oznet.ksu.edu |
| Kentucky | 606-257-7596 | jstringer@ca.uky.edu |
| Louisiana | 504-388-4087 | fowler@agctr.lsu.edu |
| Maine | 207-581-2892 | tnelson@umce.umext.maine.edu |
| Maryland | 301-432-2767 | jk87@umail.umd.edu |
| Massachusetts | 413-545-2943 | dbk@forwild.umass.edu |
| Michigan | 517-355-0096 | koelling@pilot.msu.edu |
| Minnesota | 612-624-3020 | mbaughma@forestry.umn.edu |
| Mississippi | 601-325-3150 | tomm@ces.msstate.edu |
| Missouri | 573-882-4444 | ——— |
| Montana | 406-243-2773 | efrsl@forestry.umt.edu |
| Nebraska | 402-472-5822 | fofw093@unlvm.unl.edu |
| Nevada | 702-784-4039 | walker@scs.unr.edu |
| New Hampshire | 603-862-4861 | karen.bennett@unh.edu |
| New Jersey | 732-932-8993 | vodak@aesop.rutgers.edu |
| New Mexico | 505-827-5830 | frossbach@state.nm.us |

| | | |
|---|---|---|
| New York | 607-255-4696 | pjs23@cornell.edu |
| North Carolina | 919-515-5574 | hamilton@cfr.cfr.ncsu.edu |
| North Dakota | 701-231-8478 | mjackson@ndsuext.nodak.edu |
| Ohio | 614-292-9838 | heiligmann.1@osu.edu |
| Oklahoma | 405-744-6432 | cgreen@okway.okstate.edu |
| Oregon | 541-737-3700 | forestre@ccmail.orst.edu |
| Pennsylvania | 814-863-0401 | fj4@psu.edu |
| Rhode Island | 401-874-2912 | tom@uriacc.uri.edu |
| South Carolina | 864-656-4851 | gsabin@clemson.edu |
| South Dakota | 605-688-4737 | jball@doa.state.sd.us |
| Tennessee | 423-974-7346 | wclatterbuck@utk.edu |
| Texas | 409-273-2120 | a-dreesen@tamu.edu |
| Utah | 801-797-4056 | mikek@ext.usu.edu |
| Vermont | 802-656-4280 | tmcevoy@together.net |
| Virginia | 540-231-7679 | jej@vt.edu |
| Washington | 509-335-2963 | baumgtnr@wsu.edu |
| West Virginia | 304-293-2941 | tpahl@wvu.edu |
| Wisconsin | 608-262-0134 | ajmartin@facstaff.wisc.edu |
| Wyoming | 307-766-3103 | ——— |

# Appendix B. Attorneys General in the United States

| | | |
|---|---|---|
| Alabama | 334-242-7300 | www.e-pages.com/aag |
| Alaska | 907-465-3600 | www.law.state.ak.us |
| Arizona | 602-542-4266 | www.azag.com |
| Arkansas | 501-682-2007 | www.state.ar.us/ag/ag.html |
| California | 916-324-5437 | www.caag.state.ca.us |
| Colorado | 303-866-3052 | www.state.co.us/gov_dir/dol/index.htm |
| Connecticut | 860-566-2026 | www.cslnet.ctstateu.edu/attygenl |
| Delaware | 302-577-3838 | www.state.de.us/govern/elecoffl/attgen/agoffice.htm |
| Florida | 904-487-1963 | legal.firn.edu |
| Georgia | 404-656-4585 | www.ganet.org/ago |
| Hawaii | 808-586-1282 | www.state.hi.us/ag |
| Idaho | 208-334-2400 | www.state.id.us/ag/homepage.htm |
| Illinois | 312-814-2503 | www.acsp.uic.edu/~ag |
| Indiana | 317-233-4386 | www.ai.org/hoosieradvocate/index.html |
| Iowa | 515-281-3053 | www.state.ia.us/government/ag/index.html |
| Kansas | 913-296-2215 | lawlib.wuacc.edu/ag/homepage.html |
| Kentucky | 502-564-7600 | www.law.state.ky.us |
| Louisianna | 504-342-7013 | www.laag.com |
| Maine | 207-626-8800 | —— |
| Maryland | 410-576-6300 | www.oag.state.md.us |
| Massachusetts | 617-727-2200 | www.magnet.state.ma.us/ag |
| Michigan | 517-373-1110 | —— |
| Minnesota | 612-296-6196 | www.ag.state.mn.us |
| Mississippi | 601-359-3692 | www.ago.state.ms.us |

| | | |
|---|---|---|
| Missouri | 573-751-3321 | www.state.mo.us/ago/homepg.htm |
| Montana | 406-444-2026 | —— |
| Nebraska | 402-471-2682 | —— |
| Nevada | 702-687-4170 | www.state.nv.us/ag |
| New Hampshire | 603-271-3658 | www.state.nh.us/oag/ag.html |
| New Jersey | 609-292-4925 | www.state.nj.us/lps |
| New Mexico | 505-827-6000 | —— |
| New York | 518-474-7330 | www.oag.state.ny.us |
| North Carolina | 919-716-6400 | www.jus.state.nc.us/Justice |
| North Dakota | 701-328-2210 | pioneer.state.nd.us/ndag |
| Ohio | 614-466-3376 | www.ag.ohio.gov |
| Oklahoma | 405-521-3921 | —— |
| Oregon | 503-378-6002 | www.doj.state.or.us |
| Pennsylvania | 717-787-3391 | www.attorneygeneral.gov |
| Rhode Island | 401-274-4400 | www.sec.state.ri.us/genoff/pine.htm |
| South Carolina | 803-734-3970 | www.scattorneygeneral.org |
| South Dakota | 605-773-3215 | www.state.sd.us/state/executive/attorney/attorney.html |
| Tennessee | 615-741-6474 | —— |
| Texas | 512-463-2191 | www.oag.state.tx.us |
| Utah | 801-538-1326 | www.at.state.ut.us |
| Vermont | 802-828-3171 | —— |
| Virginia | 804-786-2071 | www.state.va.us/~oag/main.htm |
| Washington | 360-753-6200 | www.wa.gov/ago |
| West Virginia | 304-558-2021 | —— |
| Wisconsin | 608-266-1221 | www.doj.state.wi.us |
| Wyoming | 307-777-7841 | www.state.wy.us/~ag/index.html |

# Glossary

**Abstract of title.** A historic summary of property title transfers, with evidence of a current, proper title and an opinion on any encumbrances that may impair or restrict title, usually prepared by an attorney or title company.

**Acre.** An area of land containing 43,560 square feet, approximately equal to the playing area of a football field.

**Addendum.** A written statement, memorandum, change, or endorsement of the original terms of a document. An addendum to a will is known as a *codicil*, and an addendum to an insurance policy is known as a *rider.*

**Administrator.** A person appointed by the probate court to handle the affairs of a decedent who has died intestate (without a will).

***Ad valorem.*** A Latin term that literally means "at value." It applies to property assessments that are based on current fair market value of an asset and not necessarily on its potential to produce value.

**Adverse possession.** When an owner's claim to property is passed to another against his or her will by action of a public entity, through eminent domain, by action of a court judgment, or through some other process whereby another individual stakes a claim against an owner's property. State statutes define the conditions under which an individual can claim the property of another through adverse possession.

**Affidavit.** A written statement that the author claims is the truth. In most states, the author must also swear an oath before a public notary affirming the statement is true.

**Agent.** A person who acts at the behest of another who is the sole beneficiary of those actions. An agent has a fiduciary responsibility to the person (or entity) he or she represents.

**Alienation.** The process whereby a property owner passes title to another.

**Amortization.** A schedule (usually monthly) for repaying debt in equal amounts that includes interest paid on the outstanding balance. As the debt is repaid, more and more of each monthly payment is attributed to the original loan amount. Early payments are mostly interest on the principal sum.

**Annuity.** A series of equal payments to be paid over a period as specified by a contract. A life insurance annuity is an investment where the insured pays a sum of money to an insurer who agrees to provide an "annuity" beginning at some point in the future, as specified in a contract. The person who receives the annuity (which can usually also be paid as a single lump sum) is called the "annuitant." The annuitant may be the insured, or someone the insured has appointed as beneficiary.

**Arbitration.** A process whereby differences are settled between two or more parties based on the decisions of one or more persons who know the facts of the dispute. In arbitration, the disputants usually agree to accept and live by the decision of the arbitrator(s) even before the process begins.

**Assignment.** A transfer of rights from one party (assignor) to another (assignee).

**Attractive nuisance.** A feature of private property, such as a pond, that holds danger for children but attracts them nevertheless.

**Balloon note.** A loan agreement whereby the lender agrees to a long loan period to make monthly payments more affordable for the borrower, with the condition that the borrower repay the entire loan usually long before the period expires. A note or loan balloons on the date the borrower agrees to repay the full balance.

**Basal area.** A measure of tree density. It is determined by estimating the total cross-sectional area of all trees measured at breast height (4.5 feet above the ground) and expressed in square feet per acre.

**Beneficiary.** A person who is legally entitled to benefit from a trust, life insurance annuities, or the profits of an estate, without having control over the assets.

**Biodiversity.** An approach to managing land that espouses practices that maintain or enhance diversity of plants and animals.

**Bond.** Something of value, usually money, that is offered to secure a promise to fulfill an obligation. A "performance bond" is a common element of a timber sale contract.

**Breach.** Failure to fulfill the terms of a contract.

**Capital.** Any asset that can be used to generate more capital, income, or wealth.

**Capital gains.** Profits obtained from the sale of capital.

**Capital improvement.** An expenditure that increases the value or extends the useful life of a capital asset.

**Capital loss.** When a capital asset is sold for less than its original cost plus capital improvements less allowable depreciation.

**Capitalization.** A method of determining property value that considers potential net income and a rate of return on investment over a fixed period.

***Caveat emptor.*** A Latin term that means *buyer beware. Caveat venditor*, "seller beware."

**Chattel.** Any personal property.

**Closing.** The event where legal title is transferred and all financial matters and other conditions of a purchase agreement are finally resolved.

**Clouded title.** An encumbrance created by a lien, claim, or document that casts doubt on the current title.

**Codicil.** *See* addendum.

**Collateral.** Something of value that is offered to secure the repayment of a debt.

**Collusion.** An illegal agreement between two or more people to defraud someone else.

**Color of title.** A deed that appears to be valid but in fact is not, usually due to fraud, such as forgery.

**Compensatory damages.** A monetary award in a judgment that compensates the injured party for actual losses.

**Condemnation.** The result of a government's right of eminent domain to take property when it is deemed in the public's best interest to do so. A condemnation requires "just compensation" to the owner whose interests have been condemned. *See* eminent domain.

**Consequential damages.** A monetary award in a judgment that is equal to the loss the breaching party could have foreseen as a result of his or her nonperformance.

**Consideration.** Value that is offered to someone in exchange for a promise or set of promises.

**Conveyance.** The transfer of an interest in real estate.

**Covenant.** A promise or a binding agreement that can pass from one property owner to the next when written into a deed. A "Covenant Running with the Land" binds future owners to maintain the covenant, irrespective of when it was created.

**Covenant of seizen.** The grantor's promise and warranty (in a deed) that he or she does own the property and has full and exclusive rights to convey a clear and marketable title to the grantee.

**Cruise.** A survey of forest stands to determine the number, size, and species of trees, as well as terrain, soil conditions, access, and any other factors relevant to forest-management planning.

**Curtesy.** (From common law, an antiquated concept that is now governed by state statute.) A husband's preemptive rights to his decedent wife's estate. It is most apt to apply to the couple's domicile.

**Damages.** Money awarded by a court for injuries sustained as a result of the actions or inactions of another. *See* compensatory damages, consequential damages, liquidated damages, and punitive damages.

**Deed.** A legal document that describes the rights a person has to real property.

**Deed restriction.** Also known as "restrictive covenant." Any condition set forth in a deed that defines or limits the extent of a deed holder's rights.

**Defamation.** An attempt to injure another person's reputation by making false statements.

**Defendant.** A person against whom a "plaintiff" has filed a suit.

**Deposition.** Sworn testimony given outside of court, usually in preparation for trial.

**Depreciation.** As applies to IRS rules, the annual cost of an asset used for business purposes, which can also be deducted from gross income. Business equipment depreciates according to schedules established by the IRS.

**Devise.** A testamentary gift of real estate from a "devisor" to the person receiving the gift, the "devisee."

**Diameter or DBH.** Tree diameter at breast height, taken at 4.5 feet above average ground level at the base of the tree.

**Discovery.** The process of obtaining facts in connection with a lawsuit, usually in preparation for trial. Depositions are part of the discovery process.

**Dominant estate.** When the rights to real estate have been separated, the "estate" created by the separation of rights is the *dominant estate.* It is the property that benefits from an easement. The estate from which the rights have been separated is the *servient estate.* The holder of an easement has the dominant estate, while the holder of the property from which the easement evolved has the servient estate.

**Dower.** (From common law, an antiquated concept that is now governed by state statute.) A wife's preemptive rights to her decedent husband's estate. It is most apt to apply to the couple's domicile.

**Durable power of attorney.** An extreme type of agency relationship that gives someone the power to make decisions for another person even after that person has become incapacitated (but is not dead).

**Duress.** A person who is compelled or forced to do something against his or her will is acting under duress.

**Easement.** A separation (commonly irrevocable) of the rights an owner normally has in real estate that vests a portion of those rights to another person or entity is known as an easement.

**Easement appurtenant.** Also known as an *appurtenant easement,* it is a right to use another's land that is attached to and benefits the title of, most commonly, an adjacent property. For instance, an easement to a neighbor for a driveway is noted as an easement appurtenant in the neighbor's property title.

**Easement by necessity.** Usually a temporary right to use another's land for access to otherwise landlocked properties. Some states require this type of easement (from a servient estate) for defined uses, such as farming or forest management. *See* dominant estate.

**Ecosystem health.** A somewhat controversial term among foresters when applied to forests; as used here, it implies that a healthy forest ecosystem is one where the essential ecosystem integrity is maintained largely within the limits imposed by nature. The term also implies that careless or excessive human use can lead to a degradation of ecosystem integrity.

**Ecosystem management.** A new approach to managing forests that "attempts to meet human needs without disrupting the integrity of the ecological processes by which forest ecosystems sustain themselves" (David Brynn, Middlebury, Vermont).

**Eminent domain.** A right of government (local, state, and federal) to take private property for public use. *See* condemnation.

**Equitable title.** An interest in property held by someone other than the deed holder, usually created by a contract. A timber sale creates an equitable title to gain access to woodlands and harvest timber as specified in the contract.

**Escheat.** A right of state government to claim private property of a decedent who has died intestate and without legal heirs.

**Escrow.** The holding of money or property by a disinterested party until the conditions of a contract are met. By state law, some types of escrow accounts are non–interest bearing.

**Even-aged.** A situation in a forest stand where the difference in age between trees forming the main canopy usually does not exceed 20 percent of the age of the stand at "rotation."

**Exclusionary.** A clause in a contact that states, in effect, "If it is not listed here, assume the act or practice is excluded from the agreement."

**Exculpatory clause.** Although not fully recognized in all states, it is a clause in a contract that releases someone, usually a landowner, from personal liability.

**Executed contract.** When all of the conditions of a contract have been met by all parties, the contract is said to be "executed."

**Executor.** A person named in a will to act as the personal representative of a decedent.

**Express contract.** A contractual agreement specifically stated, either oral or written.

**Fiduciary.** A person with a legal duty to act for the benefit of another. An agent has a fiduciary responsibility to his or her principal.

**Finder's fee.** In forestry, an amount paid by a timber buyer to a third party for information that leads to a purchase of standing timber.

**Foreclosure.** A court-ordered process where property is sold to repay a debt on which the property holder has defaulted.

**Forest type.** A natural group or association of different species of trees that commonly occur together over a large area. Forest types are defined and named after one or more dominant species of trees in the type, such as the spruce-fir and oak-hickory types.

**Fraud.** Any illegal act or misrepresentation of facts intended to deceive or cheat another person.

**General partnership.** All members of the partnership share in management, profits, and liabilities.

**Grant.** In realty, a conveyance of interests in land by deed.

**Grantee.** The person to whom grant of a deed is made. Usually the purchaser.

**Grantor.** The person who is passing or alienating title. Usually the seller.

**Habitat.** The place where a plant or animal can live and maintain itself.

**Haul road.** An interior forest road that connects the log landing with public roads, usually constructed with gravel and designed to support the weight of fully loaded log trucks.

**High grading.** An exploitive logging practice that removes only the best, most accessible, and marketable trees in a stand.

**Holding period.** As it relates to IRS rules, it is the minimum period that a taxpayer must own a capital asset before selling it with the advantage of lower tax rates on income from long-term capital gains.

**Holographic will.** A handwritten will that is signed by the testator, but not witnessed.

**Implied contract.** A contractual obligation implied by the actions (or inactions) of the parties.

**Indemnification.** To protect against injury, loss, or damages due to the actions of another. In a timber sale contract, the buyer will usually agree to" indemnify and save harmless the seller from any injuries, losses, or damages resulting from the harvesting and transportation of timber."

**Injunction.** A court order to stop someone from doing something.

**Intestate.** A person who dies without a will leaves an estate that is *intestate.*

**Involuntary lien.** A lien imposed on property for back taxes, or for other reasons, without the consent of the owner.

**Interrogatories.** Written questions, the answers to which are also written by a person who swears the answers are complete and truthful. A pretrial method of "discovery."

**Irrevocable.** Immutable, or incapable of being changed in any way.

**Joint and several liability.** When two or more people agree to the terms of a contract, it should specify that they are all liable together (jointly), as well as independently (severally), if one or more of the parties refuses to hold up their end of the bargain.

**Judgment.** A decision rendered by a court.

**Landing.** A cleared area varying in size from a quarter-acre to 2 or more acres, usually near a public road, where logs from the woodlands are skidded and sorted, then loaded onto trucks for transport to various markets.

**Libel.** A defamatory statement in writing that cannot be proven by facts.

**Lien.** A claim imposed against the title of a property as security, such as for repayment of a debt.

**Limited partnership.** A union of individuals usually for business purposes, where some individuals share profits and losses but do not make decisions for the partnership (limited partners), while other partners run the business (general partners).

**Limited power of attorney.** A legal instrument that authorizes an individual to act as an agent of another person but with limitations on the time or scope of authority. *See* durable power of attorney and power of attorney.

**Liquidated damages.** A monetary amount agreed to by the parties of a contract that represents reasonable compensation for damages that might result if one of the parties fails to perform.

**Living will.** A properly executed and witnessed document that outlines the extent to which medical personnel should use life-support and other heroic measures to sustain the life of the person named in the will.

**Log.** A section of the main stem (trunk) of a tree, varying in length and minimum diameters according to local market standards, that is usually sawn into lumber. As a verb, log refers to the process of harvesting, extracting, and transporting logs to a mill.

**Marking guidelines.** The rules used to mark a forest stand to implement a prescription.

**Maturity.** In forest trees, it is usually expressed in two ways: (1) financial maturity—when a tree has reached the point where it has maximized value growth from the market's perspective; and (2) biological maturity—when a tree has reached the point where the energy costs of maintaining itself exceeds the energy input from photosynthesis. Biologically mature trees are usually much older than the age of financial maturity.

**MBF.** An abbreviation for "one thousand board feet."

**Merchantable.** A standing tree that has a net value when processed and delivered to a mill or some other market after all harvesting costs, transportation costs, and profit are considered.

**Mineral rights.** A portion of the bundle of rights that allows the holder to explore and extract any minerals (or specific minerals) at any time in the future.

**Monument.** A fixed object identified by a surveyor, usually for the purposes of marking property corners.

**Motion.** A formal request of the court during a legal proceeding.

**Notary (or Notary Public).** A public official who will witness a signature and verify that the person is, in fact, who he or she claims to be. When a document is notarized, it includes an imprint of the notary's seal.

**Partnership.** An agreement between two or more people, usually for the purposes of doing business together. *See* general partnership and limited partnership.

***Per capita.*** A Latin term used in trusts and wills that literally means "by heads." A division of property *per capita* means equal shares to all.

***Per stirpes.*** A Latin term used in trusts and wills that means "by stocks, or by roots." A division of property *per stirpes* means a parent's share of an estate is divided among his or her children. If Mom dies before Grandpa, Mom's share of Grandpa's estate is divided among her children.

**Perjury.** Lying under oath.

**Personal representative.** *See* executor.

**Plaintiff.** The person who initiates a lawsuit against a defendant.

**Power of attorney.** A legal instrument that authorizes an individual to act as an agent of another person. *See* limited power of attorney and durable power of attorney.

**Prescription.** A course of action to effect change in a forest stand.

**Principal.** In an agency, the principal is a person who has authorized an agent to act on his or her behalf. In finance, principal is an amount originally invested or the outstanding balance (excluding interest) of a loan.

**Probate.** The legal process of proving the validity of a decedent's will and dispersing the estate to legal heirs.

**Punitive damages.** A monetary award in a court judgment that is in addition to compensatory and consequential damages, for the purposes of punishing the defendant.

**Ratify.** To accept or approve of the acts of another person after the fact.

**Recording.** The process of filing documents with the town or county clerk for the purpose of giving public notice of claims, and establishing priority over subsequent claims.

**Regeneration.** The natural or artificial (by planting) renewal of trees in a stand.

**Restrictive covenants.** Written conditions imposed on a deed that prohibit or restrict certain activities or uses of the land. Any condition that reduces the full bundle of rights is a restrictive covenant that a grantee agrees to honor and maintain. *See* covenants that run with the land.

**Retainer.** A fee paid to a service provider that is an advance on the cost of future services. In forestry, a retainer can be identified as "consideration" from the "principal" in a contract that establishes an agency relationship with a consulting forester. A retainer paid to a lawyer is usually an advance on the cost of future services.

**Rider.** An amendment to a document that applies especially to property insurance. *See* addendum.

**Right-of-first-refusal.** A form of "consideration" in a contract, where one party agrees to consider a future offer from the other party before soliciting other offers. A right-of-first-refusal is often an element in a timber sale contract where the buyer, or the buyer's agent, also provides services.

**Right-of-way.** A type of easement where a titleholder gives or sells (or leases) the right to cross over a portion of his or her land. The term is used commonly for roadways, gas and water mains, and sewer, power, and telephone lines.

**Riparian rights.** Rights and obligations that relate to land abutting water, water courses, and the use of water, especially in states west of the Mississippi.

**Rotation.** Usually considered to be the length of time it takes to grow an even-aged stand to the point of financial maturity. *See* maturity.

**Royalty.** A share of profits claimed by the owner of property who allows another to harvest, recover, or use assets from the property.

**Sawlog.** *See* log.

**Sawtimber.** Trees that have obtained a minimum diameter at breast height that can be felled and processed into sawlogs.

**Servient estate.** The property from which an easement has been granted. *See* dominant estate.

**Severance damages.** Usually a sum of money paid to a property owner when a condemnation by eminent domain of a portion of his or her property causes the remaining property to lose value.

**Silviculture.** The "art and science" of growing forests for timber and other values.

**Skidder.** A 4-wheel-drive tractorlike vehicle, articulated in the middle for maneuverability, with a cable or grapple on the back end for hauling logs from stump to a landing area, where they are loaded onto trucks.

**Skid trail.** Any path in the woods over which multiple loads of logs have been hauled. A trail that enters a main log landing area is called a "primary skid trail."

**Slander.** A defamatory oral statement that cannot be proven by facts.

**Stand.** A community of trees occupying a specific area and "sufficiently uniform in composition, age, arrangement, and condition as to be distinguishable from the forest on adjacent areas."

**Statute of limitations.** A state law that sets limits on the amount of time a plaintiff has to file suit against a defendant.

**Stipulation.** An agreement, promise, or guarantee. Also a condition or requirement of a legal document. In court, it is a mutual agreement as to the source and validity of information presented during a legal proceeding.

**Stumpage.** The value of timber as it stands in the woods just before harvest (on the stump). In some parts of the country it is considered to be a residual sum; that is, the market value of rough lumber minus all the costs of production back to the stump. Whatever is leftover is "stumpage," the value of a standing tree.

**Sublease.** An assignment of a lessee's rights and responsibilities to another person.

**Subpoena.** A written order from a court requiring an individual to appear and testify.

**Sustainability.** The capacity of an ecosystem to provide benefits in perpetuity without substantially compromising ecosystem integrity.

**Title.** The body of facts or events that support a claim of ownership in real or personal property. In real estate, a deed, if unclouded and proper, is evidence of title. *See* deed.

**Title insurance.** A type of insurance policy that is fully paid for when title passes to a new owner. It warrants and guarantees the title is valid, unclouded, and marketable.

**Tort.** A noncontractual, civil wrong that results in injury to another person. Stealing from someone is a tort.

**Trespass.** An uninvited use or encroachment of a person's rights to private property.

**Trust.** An agreement or contract where the legal and beneficial interests in property are separated according to the terms of the document.

**Trustee.** The person or entity that holds and manages property (i.e., controls the legal interests of the property) for the exclusive benefit of another, known as the "beneficiary."

**Trustor (or Settlor).** The person who created a trust and whose property is deeded to the trust.

**Usury.** Extracting a rate of interest on a loan that is excessive, as defined by state statute.

**Variance.** An exception to current rules for good cause. In real estate, variances apply to zoning regulations.

**Vest.** To grant or endow with rights, authority, or property. A properly alienated deed to forest land vests the right and obligations of ownership to the grantee.

**Voidable.** A contract that appears valid but is not because of errors or other flaws that cause injury to one of the parties. The injured party's claim of a "voidable" contract may cause a court to find the contract (or parts of it) void.

**Will.** A properly executed and witnessed document drawn up by an individual providing for the disposition of real and personal property upon the person's death.

**Yield.** Total forest growth over a specified period of time, less mortality, unmarketable fiber, and cull.

**Zoning.** Local ordinances that control how land is used in a municipality, especially as it relates to building. *See* variance.

# References and Suggested Readings

Ambrose, S. E. 1996. *Undaunted Courage.* Simon and Schuster, New York. 521 pp.

American Forest and Paper Association. 1994. *Federal Laws and Regulations Affecting Private Forestry: What Foresters and Landowners Need to Know to Be in Compliance.* American Forest and Paper Association, 1111 19th St., N.W., Washington, DC 20036.

Barlowe, Raleigh. 1990. *Who Owns Your Land?* Publication No. 126. Southern Rural Development Center, Mississippi State, MS 39762. 9 pp.

Barry, Vincent. 1994. "Moral Issues in Business." In *Ethics in Forestry.* Timber Press, Portland, OR.

Birch, T. W. 1994. *The Private Forest: Land Owners of the United States, 1994 (Preliminary Findings).* White paper. USDA Forest Service, Northeastern Forest Experiment Station. 37 pp.

Bove, Alexander A., Jr. 1991. *The Complete Book of Wills and Estates.* Henry Holt and Co., New York. 268 pp.

Brodsky, S. L. 1991. *Testifying in Court: Guidelines and Maxims for the Expert Witness.* American Psychological Assoc., Washington, DC. 208 pp.

Brown, Curtis. M. 1981. *Evidence and Procedures for Boundary Location.* 2d ed. John Wiley and Sons, New York. 450 pp.

Frascona, Joseph L., Edward J. Conry, Gerald R. Ferra, Terry L. Lantry, Bill M. Shaw, George J. Siedel, George W. Spiro, and Arthur D. Wolfe. 1984. *Business Law Text and Cases: The Legal Environment.* 2d ed. W. C. Brown, Publishers, Dubuque, IA. 1009 pp.

Gregory, G. R. 1972. *Forest Resource Economics.* The Ronald Press, New York.

Hazard, G. C., and M. Taruffo. 1993. *American Civil Procedure.* Yale University Press, New Haven, CT. 230 pp.

Irland, Lloyd C., ed. 1994. *Ethics in Forestry.* Timber Press, Portland, OR. 458 pp.

Kaplan, E. J. 1979. *Evidence.* C. C. Thomas, Publisher, Springfield, IL. 348 pp.
Kaufman, P. C., and S. H. Green. 1987. *Understanding Estate Planning and Wills.* 2d. ed. Longmeadow Press, Stamford, CT. 102 pp.
Lusk, Harold F., Charles M. Hewitt, John D. Donnell, and A. James Barnes. 1978. *Business Law: Principles and Cases.* 4th ed. R. D. Irwin, Inc., Homewood, IL. 1365 pp.
Martus, C. E., Harry L. Haney, and William C. Siegel. 1995. "Local Forest Regulatory Ordinances." *Journal of Forestry* 93(6): 27–31.
McEvoy, Thom J. 1995. *Introduction to Forest Ecology and Silviculture.* University of Vermont Extension Service Brieflet No. 1387. Burlington, VT. 75 pp.
McLauchlan, W. P. 1977. *American Legal Process.* John Wiley and Sons, New York. 218 pp.
Pearson, Karl G., and Michael P. Litka. 1980. *Real Estate: Principles and Practices.* 3d ed. Grid Publishing Co., Columbus OH. 347 pp.
Phillips, D. T., and B. S. Wolfkiel. 1994. *Estate Planning Made Easy.* Dearborn Financial Publishing, Chicago, IL. 227 pp.
Prestopino, C. J. 1989. *Introduction to Estate Planning.* Dow Jones–Irwin, Inc., New York. 557 pp.
Ring, A. A., and J. Dasso. 1977. *Real Estate Principles and Practices.* 8th ed. Prentice-Hall, Englewood Cliffs, NJ. 715 pp.
Schneider, A. E., et al. 1967. *Understanding Business Law.* 4th ed. McGraw-Hill, New York. 502 pp.
Schneider, Arnold E, John E. Whitcraft, R. Robert Rosenberg, and Robert Olaf Skar. 1967. *Understanding Business Law.* 4th ed. McGraw-Hill, New York. 502 pp.
Sharpe, Grant W., Clare W. Hendee, Wenonah F. Sharpe, and John C. Hendee. 1995. *Introduction to Forest and Renewable Resources.* 6th ed. McGraw-Hill, New York. 664 pp.
Shumate, W. A. 1995. "Preserve Timberland and Save Taxes Through Family Partnerships." *Forest Farmer* (November–December 1995): 17–18.
Siegal, W. C., et al. 1995. *Forest Owners' Guide to the Federal Income Tax.* USDA Forest Service, Agricultural Handbook No. 708. Available from Superintendent of Documents, U.S. Government Printing Office, Mail Stop SSOP, Washington, DC 20402-9328. Phone (202) 783-3238.
Siegal, William C, William L. Hoover, Harry L. Haney, Jr., and Karen Liu. 1995. *Forest Owners' Guide to the Federal Income Tax.* USDA Forest Service, Agricultural Handbook No. 708. Available from Superintendent of Documents, U.S. Government Printing Office,

Mail Stop SSOP, Washington, DC 20402-9328. Phone (202) 783-3238.

Sigler, J. A. 1968. *An Introduction to the Legal System.* The Dorsey Press, Homewood, IL. 248 pp.

Small, S. J. 1992. *Preserving Family Lands: Essential Tax Strategies for the Landowner.* 2d ed. Landowner Planning Center, Boston, MA. 99 pp.

Smith, David M., Bruce C. Larson, Matthew J. Kelty, P. Mark, and S. Ashton. 1997. *The Practice of Silviculture.* 9th ed. John Wiley and Sons, New York. 537 pp.

Terry, Brent W. 1995. *The Complete Idiot's Guide to Protecting Yourself from Everyday Legal Hassles.* Alpha Books, New York. 309 pp.

Zhang, D. 1996. "State Property Rights: What, Where and How?" *Journal of Forestry* 94(4): 10–15.

# Index